Large Aperture Array Radar Systems
for Automotive Applications

Fabian Schwartau

Large Aperture Array Radar Systems for Automotive Applications

with 60 figures

Large Aperture Array Radar Systems for Automotive Applications

Von der Fakultät für Elektrotechnik, Informationstechnik, Physik
der Technischen Universität Carolo-Wilhelmina zu Braunschweig

zur Erlangung des Grades eines Doktors
der Ingenieurswissenschaften (Dr.-Ing.)

genehmigte Dissertation

von Fabian Valentin Schwartau

aus Hamburg

eingereicht am: 12.01.2021

mündliche Prüfung am: 05.08.2021

1. Referent: Prof. Dr.-Ing. Jörg Schöbel
2. Referent: PD Dr.-Ing. habil. Robert Geise

Druckjahr: 2021

Bibliografische Information der Deutschen Nationalbibliothek
Die Deutsche Nationalbibliothek verzeichnet diese Publikation in der Deutschen Nationalbibliografie; detaillierte bibliographische Daten sind im Internet über http://dnb.d-nb.de abrufbar.
1. Aufl. - Göttingen: Cuvillier, 2021
Zugl.: (TU) Braunschweig, Univ., Diss., 2021

Revision 258 – 9th October 2021, 13:08:09

Nonnenstieg 8, 37075 Göttingen
Telefon: 0551-54724-0
Telefax: 0551-54724-21
www.cuvillier.de

1. Auflage, 2021
Gedruckt auf umweltfreundlichem, säurefreiem Papier aus nachhaltiger Forstwirtschaft.

ISBN 978-3-7369-7507-1
eISBN 978-3-7369-6507-2

Abstract

The automotive industry is pushing towards highly assisted and even autonomous driving cars. To gather a more precise and reliable representation of the car's surroundings, the sensors and the signal processing are improving over time and are a subject to continuous research. One essential sensor is the radar, which is robust and reliable even in harsh environmental conditions. The primary downside of a radar is its low resolution compared to lidar or camera-based systems.

To mitigate these drawbacks the resolution of radar systems has to be improved. The bandwidth has to be increased to improve the range resolution, and the aperture has to be increased to improve the angular resolution.

Primarily caused by the automotive industry, fully integrated radar on chip solutions are now available and allow the construction of more complex radar systems. These radar on chip devices lay the foundation for radars that fulfill the requirements of increased resolution for future systems.

Although this work is focused automotive applications, most ideas, concepts, and calculations are also applicable to other fields. Similar systems may be used in the security sector, quality control in industrial processes, or gesture detection, to name a few examples.

This thesis shows the development of a conceptual future radar system for automotive applications. The focus is on providing a large antenna aperture to achieve the required high angular resolution. Two genetic algorithms are developed to optimize the antenna array for a good side lobe level while providing high angular resolution.

Two demonstrators are built to implement certain aspects of the proposed radar system and prove the general concept viable. The first demonstrator features a large aperture with a limited side lobe level and is using a modular approach. The modules are synchronized with a radio over fiber system. The second demonstrator uses the previously proposed antenna array, which is implemented with a synthetic aperture radar approach. The system's capabilities in a real scenario are demonstrated, and the reconstruction of a high-resolution three-dimensional image from the captured data is shown.

As larger arrays for radar systems automatically come with higher manufacturing tolerances, larger thermal expansion, and other negative impacts on the actual antenna positions, this work also analyzes the impact of such imperfections. Additionally, an algorithm is proposed, which can estimate the actual antenna displacement during operation. Therefore, the system is capable of compensating for most of the negative effects.

Kurzfassung

Die Autoindustrie strebt nach erweiterten Assistenzsystemen und sogar autonom fahrenden Autos. Um eine präzise und zuverlässige Darstellung der Fahrzeugumgebung zu erhalten, werden die Sensoren und die Signalverarbeitung stetig weiter entwickelt und sind Gegenstand aktueller Forschung. Ein zentraler Sensor ist das Radar, was auch unter erschwerten Bedingungen robust und zuverlässig ist. Der primäre Nachteil eines Radars ist die geringe Auflösung im Vergleiche zu System auf Basis von Lidar oder Kameras.

Um diese Nachteile auszugleichen, muss die Auflösung der Radarsysteme verbessert werden. Dafür muss die Bandbreite erhöht werden, um die Entfernungsauflösung zu verbessern, und die Apertur muss vergrößert werden, um die Winkelauflösung zu verbessern.

In erster Linie durch die Automobilindustrie angetrieben, stehen nun voll integrierte Radar-on-Chip-Lösungen zur Verfügung, die die Konstruktion komplexerer Radarsysteme ermöglichen. Diese Radar-on-Chip-Lösungen bilden die Grundlage für Radarsysteme, die die Anforderungen an eine erhöhte Auflösung für zukünftige Systeme erfüllen.

Obwohl sich diese Arbeit auf die Anwendung im automobilen Umfeld konzentriert, sind die meisten Ideen, Konzepte und Berechnungen auch auf andere Bereiche anwendbar. Ähnliche Systeme können z.B. im Sicherheitsbereich, bei der Qualitätskontrolle in industriellen Prozessen oder bei der Gestenerkennung eingesetzt werden, um nur einige Beispiele zu nennen.

Diese Arbeit zeigt die Entwicklung eines konzeptionellen zukünftigen Radarsystems für automobile Anwendungen. Der Schwerpunkt liegt auf der Umsetzung einer großen Antennenapertur, um die erforderliche hohe Winkelauflösung zu erreichen. Zwei evolutionäre Algorithmen werden vorgestellt, um das Antennen-Array auf einen guten Nebenkeulen-Pegel zu optimieren und gleichzeitig eine hohe Winkelauflösung zu erreichen.

Zwei Demonstratoren werden gebaut, um bestimmte Aspekte des vorgeschlagenen Radarsystems zu implementieren und die Durchführbarkeit des allgemeinen Konzepts zu zeigen. Der erste Demonstrator weist eine große Apertur mit einem begrenzten Nebenkeulen-Niveau auf und verwendet einen modularen Ansatz. Die Module sind mit einem Radio-over-Fiber-System synchronisiert. Der zweite Demonstrator verwendet die zuvor entworfene Antennenanordnung, die mit einem Radar mit synthetischer Apertur realisiert wird. Die Fähigkeiten des Systems werden in einem realen Szenario demonstriert und die Rekonstruktion eines hochauflösenden dreidimensionalen Bildes aus den erfassten Daten gezeigt.

Da größere Arrays für Radarsysteme automatisch mit höheren Fertigungstoleranzen, größerer thermischer Ausdehnung und anderen negativen Auswirkungen auf die tatsächlichen Antennenpositionen einhergehen, werden in dieser Arbeit auch die Auswirkungen solcher Störungen analysiert. Zusätzlich wird ein Algorithmus entwickelt, der die tatsächliche Antennenverschiebung während des Betriebs abschätzen kann. Das System in daher in der Lage, die meisten der negativen Effekte zu kompensieren.

Contents

Abbreviations

ACC adaptive cruise control
ADC analog to digital converter
AoA angle of arrival
ASIC application-specific integrated circuit
BGA ball grid array
CAN controller area network
CFAR constant false alarm rate
CS chirp sequence
CSI camera serial interface
CW continuous wave
DFT discrete Fourier transformation
FFT fast Fourier transformation
FIFO first in first out
FMCW frequency modulated continuous wave
FNBW first null beam width
FPGA field programmable gate array
HPBW half power beam width
IC integrated circuit
LD laser diode
LDC laser diode controller
LED light emitting diode
LVDS low voltage differential signal
µC microcontroller
MIMO multiple input multiple output
MRLA minimum redundancy linear array
MZM Mach-Zehnder modulator
PC personal computer
PCB printed circuit board
PD photo diode
PEC perfectly electric conductor
PID proportional-integral-differential
PN pseudo noise
RCS radar cross section
RF radio frequency
RMS root mean square
RoC radar on chip
RoF radio over fiber
Rx receiver
SAR synthetic aperture radar
SBR shooting and bouncing rays
SLL side lobe level
SMA SubMiniature version A
SNR signal to noise ratio

SPI serial peripheral interface
TCP transmission control protocol
TEC temperature controller
Tx transmitter
UART universal asynchronous receiver transmitter
USB universal serial bus

1 Introduction

The term radar stands for *radio detection and ranging* and describes the general method of measuring parameters of a target like its distance or velocity with electromagnetic waves. According to the Radio Regulations Articles of the ITU, Electromagnetic waves are located in the radio frequency range, defined as "frequencies arbitrarily lower than 3000 GHz, propagated in space without artificial guide" [1]. The foundation of radar was laid in 1887 by Heinrich Hertz, who experimentally proved that electromagnetic waves are reflected from metal surfaces [2]. The first experiments for locating objects were carried out by Christian Hülsmeyer, who registered a patent in 1904 for a radar to detect ships [3].

A radar system transmits a radio wave signal and receives the echoes from all objects' reflections in range. The radar then analyzes the received echo, for example, by measuring the delay between the transmission of the signal and reception of the echo. This allows the distance estimation to the object because an electromagnetic wave always travels with the speed of light in vacuum.

To measure the delay of a signal, it needs to be modulated. The simplest form of modulation is to use a short pulse at a specific center frequency. A plot of the returned energy over time gives a simple view of the reflections as a function of their distance. A non-modulated (i.e., continuous wave (CW)) radar cannot measure the distance to an object. As the signal is periodical, it is impossible to distinguish between the different periods. However, if the target moves, it changes the reflected signal frequency. This effect is known as the Doppler effect, which allows an estimation of the targets' velocity based on the observed frequency shift. More sophisticated radar systems are capable of measuring both distance and velocity with a more complex modulation scheme. Commonly used are, for example, frequency ramps or noise modulation.

The radar becomes even more sophisticated when it can create a radio frequency beam and point it in multiple directions. Such a radar can determine the direction of the object, allowing a location in three-dimensional space. Simple radar systems use large antennas or antenna arrays with a narrow beam, which are steered mechanically. Although a robust and straightforward design, the steering is relatively slow. The next step is to use an antenna array and steer the beam electronically using, for example, phase shifters in each path [4, pp. 1]. By choosing the correct phase of each antenna element, which compensates the path difference in a specific direction, the beam's direction can be controlled. The signal is then added and fed to the receiver. In the case of an electronically steered transmitter, the signal is split, phase-shifted, and fed to each antenna element of the transmit array. Although the steering is much faster with this concept, the larger amount of electronic components makes it much more complex and potentially more expensive.

The downside of both of these steering types is that the radar can look only in a single direction at any given time. An even more complex system is used to solve this issue, which is becoming more and more popular in recent years. It is called digital beamforming and is essentially the same as an electronically steered antenna. The vital difference is that the steering of the beam is carried out digitally on a computer. It can be applied to transmitters and receivers by using one transmitter or receiver unit per antenna element. In the receiver's case, all signals are recorded simultaneously, and the steering can be implemented in software, which allows to point the radar's beam in multiple directions at the same time. Transmit beamforming allows to digitally steer the beam into a specific direction, like in a phased array, but the illumination of multiple directions is also possible. This is also implemented by changing the phase values or delays for each transmitted signal to compensate for the path difference for the desired angle.

If the transmitters use orthogonal signals, which do not form a beam in a particular direction, it is possible to separate the different transmitters in each received signal. This type is called a multiple input multiple output (MIMO) radar and increases the spatial resolution by measuring each combination of transmitter and receiver [3]. The beamforming is carried out subsequently on a computer, allowing to steer both transmitter and receiver after the measurements are carried out. The downside of the beamforming and MIMO approaches is the even higher system complexity and cost as it requires a complete transmitter or receiver for each antenna.

With beamsteering or beamforming, it is possible to build a three-dimensional representation of the surrounding scenario. When including the Doppler information, such a radar can distinguish targets in four dimensions, and the time may be considered the fifth dimension. To get a more and more precise representation of the radar's scenario, a higher resolution is required. This primarily refers to the angular and range resolution. As will be discussed later, a higher range resolution requires a higher bandwidth of the transmitted signal, while higher angular resolution requires a larger aperture of the antenna array (relative to the wavelength). To achieve both while keeping a compact form factor, radar systems have been pushing to higher frequencies. Higher frequencies and the demand for cheaper solutions lead to increased integration of such systems, turning fully integrated radar solution into a mass product.

Especially the automotive industry has been pushing the integration of radar systems in the past years. The goal of assisted and even autonomously driving cars has led to increased demand for reliable and cheap radar solutions. Also, the demands in terms of the resolution are continuously increasing as the car's precise knowledge of its surroundings is a key component for assisted and autonomous driving. Although such cars are additionally equipped with lidar[1] and camera systems, radar systems are considered more reliable and cheaper [5,6]. Lidar and camera systems are, for example, heavily susceptible to dirt, ice or snow on the sensor itself or in the air. Table 1.1, taken from [6], summaries the different aspects of the three technologies.

The three systems, radar, lidar, and vision, are based on the concept of detecting electromagnetic waves and estimating their angle of arrival (AoA). The two differences are that radar and lidar emit their own coherent electromagnetic wave, while the vision-based system uses incoherent illuminators of opportunity, and radar is working at a much lower frequency compared to lidar and vision and has, therefore, a much lower bandwidth.

Besides the angular resolution, the lack of color and contrast is the only other negative point of a radar system. The collection of information, like the color or contrast of a target in a visual system, is used to classify the target further. Although a radar cannot measure any color or contrast information, modern systems are capable of measuring other object parameters that help to classify the object. There has been much research regarding radar systems capable of measuring reflections of different polarizations. A target's amplitude and

[1] Light detection and ranging, similar to a radar system, but it uses lasers instead of radio waves to measure the distance to a target.

Table 1.1: Comparision of different automotive sensing systems, taken from [6]

Sensor	Radar	Lidar	Vision
Range	++	+	++
Range resolution	+	++	o
Angular resolution	o	++	+
Works in bad weather	++	o	-
Works in dark	++	++	- -
Works in bright	++	+	+
Color/contrast	- -	- -	++
Radial velocity	++	o	-

polarization properties can be used for further classification, providing similar information like color or contrast [7, 8], compensating the system's drawbacks.

Compared to the other systems, the only major drawback of a radar is the low angular resolution. This is the starting point of this thesis, and the goal is to provide additional insight on how to overcome the issue of limited angular resolution in automotive radar applications.

1.1 Organization of the Thesis

This thesis covers a wide range of topics surrounding the theoretical description, the design, the simulation, the construction, and the evaluation of an automotive radar system. This introduction additionally describes the contributions of this work, followed by the notation used throughout it.

The second chapter introduces the required fundamentals for this thesis. It contains descriptions for frequency modulated continuous wave (FMCW) and chirp sequence (CS) radars, the definition of far-field conditions, algorithms for beamforming in the near- and farfield, description of array parameters, the concept of synthetic and virtual apertures, calculation of radar resolution parameters, the constant false alarm rate (CFAR) algorithm for detection of targets within the digital domain, and a few other things.

The third chapter introduces the concept of a modular radar system for automotive applications. The state-of-the-art in current systems is summarized, followed by defining goals for future radar systems. The remainder of the chapter primarily deals with the antenna array design using modified genetic algorithms for the optimization of antenna element positions and weights within the array. The proposed array is analyzed in terms of angular resolution, directivity, and an exemplary application is used to estimate the expected performance in terms of maximum range.

Three aspects of the prosed radar system are then analyzed in the fourth chapter. These aspects are the general concept of a radar system with high angular and range resolution, the usage of individual and flexible modules, and the coherent synchronization of such modules. These goals are analyzed by the design, construction, and evaluation of a demonstrator radar system. The chapter deals with the concept, the involved hardware, the calibration, the software, and the signal processing of the demonstrator. Simulations and measurements showcase the capabilities of the system.

The fifth chapter describes a second demonstrator's construction, which uses the antenna array proposed in the third chapter. It utilizes a virtual and synthetic aperture to keep the system complexity and cost low, thus having only a single transmitter and receiver yet providing the same results as the fully populated antenna array. The second demonstrator is used to capture a more realistic automotive scenario and provides a three-dimensional image.

Chapter six analyzes potential sources of manufacturing tolerances and other effects that influence the antennas' position within the array. The magnitude of the identified causes is estimated, and the effect on the array is analyzed. Subsequently, an algorithm is introduced and evaluated to measure and compensate for the displacement of individual modules of the systems.

A conclusion of the work with a summary of the results presented in the previous chapters is given in chapter seven. Additionally, it contains an outlook of the next steps towards the next generation of high-resolution radar systems. Chapter eight contains appendices.

1.2 Contributions

This thesis contributes the following additional knowledge, techniques, and analysis to the current state of the art:

- **Fundamentals for Signal Processing and Analysis of Radar Systems**
  The second chapter provides a comprehensive overview of the mathematical fundamentals required for the analysis of and signal processing for radar systems. This overview focuses on large aperture radar systems utilizing FMCW or CS modulation, but most of it is also applicable to other systems.
- **Generation Algorithm for Large Aperture Arrays**
  Two customized genetic algorithms are proposed in Chapter 3, which are capable of optimizing the positions and weights of antenna elements in a large aperture MIMO array. The algorithms additionally allow to group transmitters and receivers into modules, limiting the potential positions they can be placed at. The algorithms provide multiple parameters to optimize for different aspects of the array, like the side lobe level or the main lobe's shape.
- **Modular, Optically Synchronized, and Large Aperture Radar**
  Two radar demonstrators are developed and analyzed in Chapter 4 and 5. Together the two demonstrators showcase a future automotive radar system, which is based on a modular design with high angular resolution. The modules operate fully coherently due to the synchronization with a radio over fiber system.
- **Antenna Position Displacement Compensation**
  Chapter 6 proposes a novel algorithm capable of compensating the displacement of antenna modules within a coherent radar system during operation. All common changes in antenna positions during operation or tolerances during production can be measured and subsequently compensated.

1.3 Notation

There are a couple of notations used throughout the following work. In general, scalars are light letters, like a or A, vectors, and matrices are represented as bold letters, like $\boldsymbol{a}$ or $\boldsymbol{A}$. The indexes x, y, and z added to a vector represent the three elements of it, pointing into the three Cartesian axes.

Figure 1.1(a) shows the standard polar coordinates used throughout this work. θ is defined as the angle downward from the z-axis, φ is the angle from the positive x-axis towards the positive y-axis. Figure 1.1(b) shows a polar coordinate system based on azimuth (γ) and elevation (δ). The azimuth is defined as the angle from the positive x-axis towards the negative y-axis. The elevation is the angle pointing upwards from the xy-plane. The x-axis is the boresight axis of the radar system; antenna elements are thus placed in the yz-plane. The y-axis points to the left when looking from behind the radar, and z is oriented upwards.

Antenna arrays can be expressed in several ways. One way is to describe the weights $\boldsymbol{w}$ in square brackets, like [1,1,1,0,1]. This expression represents an array with equidistant elements. In the example, only the first three and the last one are populated. Weights can also be used to express weights for a taper. Especially for large arrays, it is more convenient to describe the distances between two elements, while an exponent can indicate the repetition of a certain distance. Curly brackets surround this type. For example, $\{1^2,2\}$ represents the same array as the one before. A distance of 1 usually translates to $\lambda/2$, if not otherwise noted.

The following operators are used:

$\|\cdot\|$	2-norm of a vector
$\lvert\cdot\rvert$	Absolute value of a scalar, vector or matrix
$\cdot^*$	Conjugate complex of scalar, vector, or matrix
$\cdot^T$	Non-conjugate transpose of a vector or matrix
$\cdot * \cdot$	Convolution of two signals
$\Re\{\cdot\}$	Real value
$\mathfrak{F}_l\{\cdot\}$	Discrete Fourier transformation on axis l
$\angle\cdot$	Phase angle of complex number
$\max\{\cdot\}$	Maximum of a vector
$\min\{\cdot\}$	Minimum of a vector
$\text{count}\{\cdot\}$	Number of elements in a vector or matrix. In case of a matrix the columns are treated as elements.
$\text{unique}\{\cdot\}$	Unique number of elements in a vector or matrix. In case of a matrix the columns are treated as elements.
$\text{mean}_l\{\cdot\}$	Mean value of a scalar, vector or matrix on axis l

All symbols used throughout this work are listed and described in the "List of Symbols" starting on page 123.

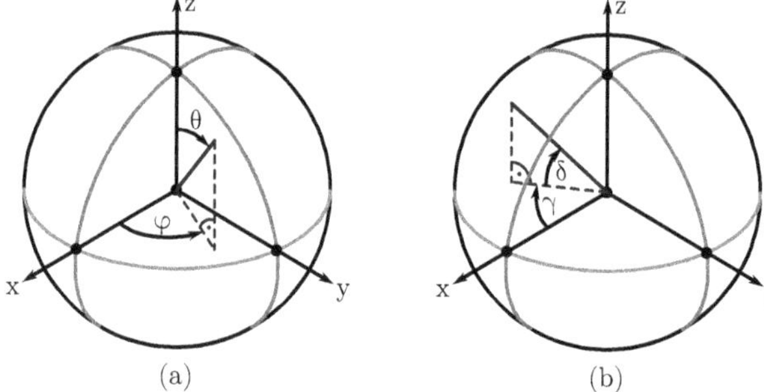

Figure 1.1: Coordinate systems in polar coordinates (a) and azimuth/elevation (b)

2 Fundamentals

This chapter contains all fundamental concepts and equations which are used throughout this work. This includes the signal representation in section 2.1, general considerations of radar systems from Section 2.2 to 2.7, beamforming and movement compensation in Sections 2.8 and 2.9, description and calculation of array parameters from Section 2.10 to 2.14, and the CFAR algorithm in Section 2.15.

2.1 Signal Representation

Signals in the time domain may be expressed as complex baseband signals [9, p. 8]. This is possible because any bandpass limited and real-valued signal $s(t)$ can be expressed by

$$s_{\mathrm{comp}}(t) = \left[s(t)e^{-j2\pi f_{\mathrm{c}}t}\right] * h_{\mathrm{LP}}(t), \tag{2.1}$$

without loosing any information, where f_{c} is the chosen center frequency (usually the center of a bandpass signal), and $h_{\mathrm{LP}}(t)$ is the impulse response of an ideal low pass filter covering the complete bandpass-bandwidth of the signal. This notation simplifies the representation of signals which are modulated to a certain carrier frequency and allows the more obvious representation of signal phase values.

The real-valued signal can be restored with

$$s(t) = 2\Re\left\{s_{\mathrm{comp}}(t)e^{+j2\pi f_{\mathrm{c}}t}\right\}. \tag{2.2}$$

As the low pass filter in (2.1) and the real part operation in (2.2) each half the resulting signal power, an additional factor of two is required in (2.2).

2.2 FMCW Radar

The FMCW radar, more precisely linear FMCW radar, in this case, modulates the frequency of the transmitted signal in linear dependency of the time. The received signal is mixed[1] with the currently transmitted signal yielding the baseband signal, which contains information about the targets it reflected on. FMCW radars can utilize up and down sweeps or a combination of both. Sawtooth modulation is a common waveform, as it allows easy calculation of range and Doppler and, thus, is used in the following description. The following equations to describe an FMCW radar are taken from [10, p. 39].

[1] A single mixer and not an IQ-mixer is assumed in the following. This leads to always positive frequencies in the baseband signal.

The delay in the received signal compared to the transmitted signal is proportional to the distance

$$\tau = \frac{2d}{c_0}, \tag{2.3}$$

d being the distance of a particular target and c_0 the speed of light.

Looking at a single sweep, the received signal has a slightly different frequency compared to the currently transmitted signal. The frequency difference, called beat-frequency f_b, depends on the distance of the target and the slope s of the frequency ramp according to

$$f_b = s\frac{2d}{c_0}. \tag{2.4}$$

The slope is defined as

$$s = \frac{B}{t_s}, \tag{2.5}$$

with the bandwidth B and the ramp duration t_s. Rearranging (2.4) and inserting the result into (2.5) gives the final equation for the target distance:

$$d = \frac{f_b t_s c_0}{2B}. \tag{2.6}$$

The observed frequency difference between the signals only corresponds directly to the beat frequency f_b if the target is not moving. If the target is moving with a radial velocity of v_r, an additional Doppler shift of

$$f_d = \frac{2v_r}{\lambda} \tag{2.7}$$

is observed. λ is the wavelength, defined as

$$\lambda = \frac{c_0}{f_c}, \tag{2.8}$$

where c_0 is the speed of light and f_c is the center frequency of the radar. The actual measured frequency is the sum or difference of those two frequencies. If the ramp is an up sweep, the measured frequency is

$$f_{bu} = f_b - f_d, \tag{2.9}$$

or

$$f_{bd} = f_b + f_d \tag{2.10}$$

in case of a down sweep.

When both up and down sweep are measured, it is possible to calculate range and Doppler. Therefore, the equations above can be rearranged to calculate the range d and radial velocity v_r of the target:

$$d = \frac{c_0 t_s}{4B}(f_{bd} + f_{bu}), \tag{2.11}$$

$$v_r = \frac{\lambda}{4}(f_{bd} - f_{bu}). \tag{2.12}$$

The maximum unambiguity range is given in [10, p. 50] as

$$d_{max} = \frac{c_0 N_{ft}}{4B}, \tag{2.13}$$

where N_{ft} is the number of samples for a single chirp (i.e., ramp or fast-time axis).

To extract the frequency information from the received baseband signal, a Fourier transformation along the time axis is commonly applied.

Figure 2.1 shows a qualitative plot of the transmitted and received FMCW radar signals in case of a single target with a positive Doppler shift, indicating a target moving towards the radar. In this case f_{bd} is higher than f_{bu}, which results in a positive radial velocity according to (2.11).

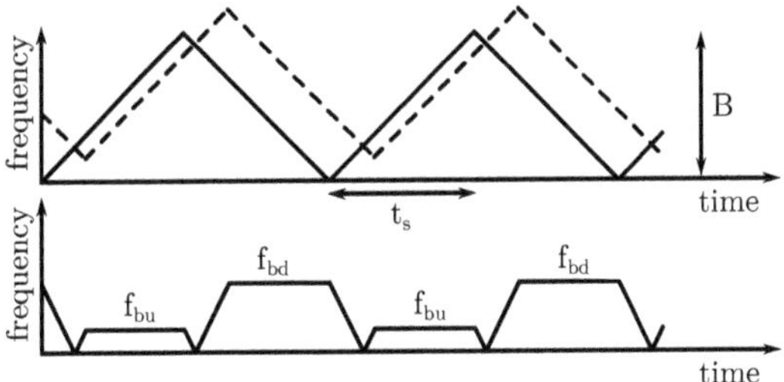

Figure 2.1: Qualitative plot of the signals in an FMCW radar, according to [10, p. 40]. The top figure shows the transmitted (solid) and received (dashed) signal. The received signal has an additional positive Doppler shift, indicating a target moving towards the radar.

Although the FMCW radar is theoretically capable of measuring range and Doppler for a target, practical implementation is hard to accomplish. To calculate both values, two measurements with an up and down sweep have to be performed. If multiple targets have been detected, they have to be matched between the two measurements. False matching will lead to false detections. This problem is why a chirp sequence radar is often preferred in applications where range and Doppler are required.

2.3 Chirp Sequence Radar

The chirp sequence modulation is very similar to the FMCW modulation but overcomes the problem of having to match two targets from two measurements to be able to calculate range and Doppler. CS also uses linear FMCW ramps, called chirps, but they are usually much shorter, multiple of them are transmitted consecutively in short order, and only up or down sweeps, but not both, are utilized. The assumption is that the target does not move more than $\lambda/4$ between two chirps.[2] It is then possible to analyze the phase change of the target over time. This is done with a Fourier transformation of a single distance over time and results in the target's radial Doppler frequency.

In practice, the measured data is put into a matrix. Each chirp fills one row of the matrix. Like in FMCW radar signal processing, a Fourier transformation along each row (called fast time axis) transforms the axis into the range axis. Every column then represents the same range over time. A second Fourier transformation along each column transforms the

[2] The additional delay between two measurements should be less than $\lambda/2$. As the wave travels to the target and back, the target may not move more than $\lambda/4$ to measure a delay of less than $\lambda/2$.

so-called slow time axis into the Doppler axis. The matrix then represents the amplitudes in dependence of range and Doppler and is called range-Doppler matrix.

The range calculation is the same as for FMCW, when based on the parameters of a single chirp, see (2.6). The same holds for the Doppler calculation from (2.7), which is applied to the slow time axis of the matrix.

A few important system parameters resulting from the chosen waveform parameters are given in [10, p. 50]. The maximum Doppler frequency is

$$f_{\mathrm{d,max}} = \frac{1}{2T_{\mathrm{s}}}, \tag{2.14}$$

with T_{s} being the period of the chirps, which is usually different to the length of the chirps t_{s}, as shown in Figure 2.2 If the Doppler frequency of a target is higher than this value, aliasing will occur and the radial velocity has a wrong value.

The maximum Doppler frequency can be converted to the maximum velocity with (2.7), yielding

$$v_{\mathrm{r,max}} = \frac{\lambda}{4T_{\mathrm{s}}}. \tag{2.15}$$

Figure 2.2 shows the qualitative plot of the transmitted and received signals in a CS radar. It is important to note that the time of a single chirp t_{s} describes the measurement's length. The transmitted chirp usually starts shortly before the first sample is taken. This ensures that the signal had enough time to return from all targets, and the sweep is stable.

When looking at a target at a distance of 150 m, the delay is already 1 µs for the return signal. In this case the chirps would typically be 2 µs to 10 µs in length. The additional delay between the start of transmission and the start of sampling is therefore important. An FMCW radar with its much longer chirps is usually not affected that much.

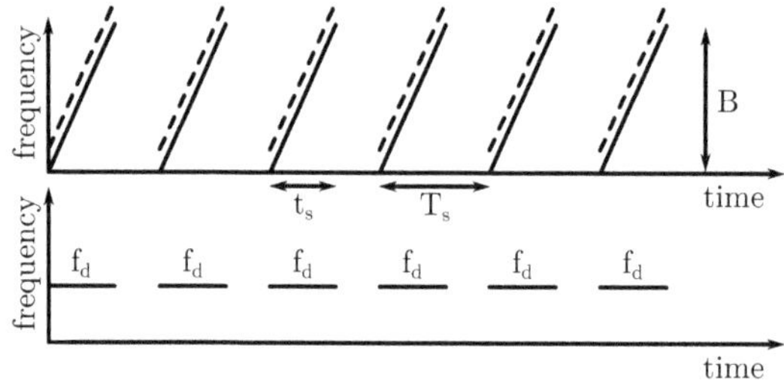

Figure 2.2: Qualitative plot of the signals in a CS radar. The top figure shows the transmitted (solid) and received (dashed) signal frequency over time. The bottom figure depicts the frequency difference between transmitted and received signal.

2.4 Radar Range Equation

The radar range equation is well known, for example, from [11, pp. 3]. It comes in various flavors depending on the known input parameters or application to different types of radar, like pulse or CW radars. The same holds for this work. Below is the standard radar range equation:

$$d < \left[\frac{P_t G_t G_r \lambda^2 \sigma}{(4\pi)^3 P_{r,min}}\right]^{1/4}. \tag{2.16}$$

P_t is the transmit power of the system, G_t and G_r describe the transmit and receive antenna gain, respectively. σ represents the radar cross section (RCS), given in square meters, and $P_{r,min}$ is the minimum detectable signal, which can be defined as

$$P_{r,min} = k_0 T_0 B_{bin} F L_{tot} \mathrm{SNR}_{min}. \tag{2.17}$$

This includes the Boltzmann constant k_0, the system temperature T_0, the receivers noise figure F, additional losses L_{tot}, which do not include the losses in the antennas (included in the antenna gain) or the losses in the receiver (included in the noise figure), and the minimum required signal to noise ratio SNR_{min}. The bandwidth B_{bin} is the resolution bandwidth after pulse compression. In the case of an FMCW or CS radar, this corresponds to the resolution in the frequency domain and is defined as

$$B_{bin} = \frac{f_{s,adc}}{N_{ft}} = \frac{1}{t_{int}}. \tag{2.18}$$

$f_{s,adc}$ is the sample rate of the analog to digital converter (ADC), N_{ft} is the number of ADC samples for one chirp or ramp, and t_{int} is the integration time of the receiver. Inserting (2.17) and (2.18) into (2.16) yields the final radar range equation for an FMCW/CS radar:

$$d < \left[\frac{P_t G_t G_r \lambda^2 t_{int} \sigma}{(4\pi)^3 L_{tot} k_0 T_0 F \mathrm{SNR}_{min}}\right]^{1/4}. \tag{2.19}$$

In many applications, there are multiple transmitters and receivers, where a few parameters can be estimated as

$$P_t = P_{t,el} N_t, \tag{2.20}$$

$$G_t = G_{t,el} N_t, \tag{2.21}$$

$$G_r = G_{r,el} N_r. \tag{2.22}$$

N_t and N_r are the number of transmit and receive elements, respectively. $P_{t,el}$, $G_{t,el}$, and $G_{r,el}$ represent the transmit power and the transmit/receive gains of an individual system element. Meaning that the total transmit power is equal to the transmit power of an individual transmitter times the number of transmitters. This applies to the gain values accordingly. Interesting is that doubling the number of transmitters corresponds to an increase of the effective field strength by a factor of four. As the number of antennas doubles, both the resulting gain and the transmit power is twice as high.

2.5 Multiple-Input Multiple-Output

According to [3, pp. 66], a MIMO radar may be defined as any radar system with multiple transmitters and multiple receivers (hence the name multiple-input multiple-output), which processes all collected information together. One of the advantages of a MIMO radar is the increased number of measurements, as any combination of transmitter and receiver is a single measurement, leading to a higher spatial resolution [3, p. 1]. This is closely related to the concept of the virtual aperture, which is introduced in Section 2.14.

A general requirement for a MIMO radar is the orthogonality of the transmitted waveforms, as every transmitter has to be separated within every received signal. A simple approach to achieve orthogonality between the transmitters is to have only one individual transmitter active at any given time. More advanced systems employ modulation types that are inherently orthogonal to each other by design, even when transmitted simultaneously. For example, one may use different pseudo noise (PN) sequences for each transmitter and use a correlation at the receiver to separate the signals again.

Although the orthogonality is only achieved by the transmitter separation in time throughout this work, the presented radars are MIMO systems.

2.6 Resolution

To achieve a more precise mapping of the surroundings of a radar, a higher resolution is required. The resolution of a radar is an important figure for its performance. There are two spatial types of resolution involved, the range resolution and the crossrange resolution.

In a radar, the range is usually measured by comparing the received with the transmitted signal and calculating a delay between the two signals. No matter how the delay is calculated, for example, by actually measuring the delay of a pulse or by correlation, the bandwidth of the transmitted signal is the primary factor for the resolution. In the case of a pulse radar, the pulse will potentially become shorter when using more bandwidth. For frequency or phase-modulated radars, the range resolution is given by

$$\Delta R = \frac{c_0}{2B}, \tag{2.23}$$

with c_0 being the speed of light and B the bandwidth of the system, according to [11, p. 21].

The crossrange or angular resolution is defined as the minimum angle between two targets, which still allows to separate the targets from each other. The angular resolution corresponds to half the first null beam width (FNBW), which is the angular distance between the two nulls on each side of the main lobe divided by two. According to [12, p. 40], half the FNBW can be approximated with the half power beam width (HPBW) of the main lobe.

The HPBW on the other hand can be estimated from the aperture of the virtual antenna array[3]. According to [4, p. 20] and [12, p. 655–656], the half power beamwidth of a fully populated linear array with no extra weighting can be approximated with

$$\text{HPBW} \approx 50.8\,^{\circ}\frac{\lambda}{D}. \tag{2.24}$$

[3]The concept of virtual antennas is introduced in Section 2.14.

Figure 2.3 shows the simulated radiation pattern[4] of a fully populated linear array of 20 isotropic elements with $\lambda/2$ spacing. Therefore, the aperture is $D = 20\lambda/2$, resulting in a beamwidth value of approximately 5.3 ° when inserted into (2.24). The simulation in 2.3 shows a beamwidth of 5.1 °, which matches the estimated value very well.

The HPBW can be calculated numerically using either the radiation pattern or the array factor. The pattern is calculated with a very high resolution, and the -3 dB point is determined on both sides of the main lobe. The angular distance between the two points is the HPBW.

It is common to apply additional taper functions to both the measured samples over time and the different antennas. These window functions will always negatively influence the resolution in both cases. When applying an additional window on the received signal along the time axis, the range resolution will become worse. The same applies to a window multiplied along the antenna elements, resulting in a worse angular resolution. Figure 2.3 shows the effect on the angular resolution in the case of a Hamming window. [10, p. 33] provides a table with various windows and the effects on the signals. Any window will increase the width of the main lobe. For example, a Hamming window will degrade the resolution (range or angular) by a factor of 1.3, independent of the side lobe level (SLL). The benefit of an additional window function is the reduced SLL, either in range or angle, depending on the axis the window is applied to. Of course, it is possible to apply windows in both axes.

The HPBW and thus the angular resolution is defined as the resolution in boresight direction, if not otherwise noted. When not looking in the boresight direction, the effective aperture will decrease and worsen the angular resolution accordingly.

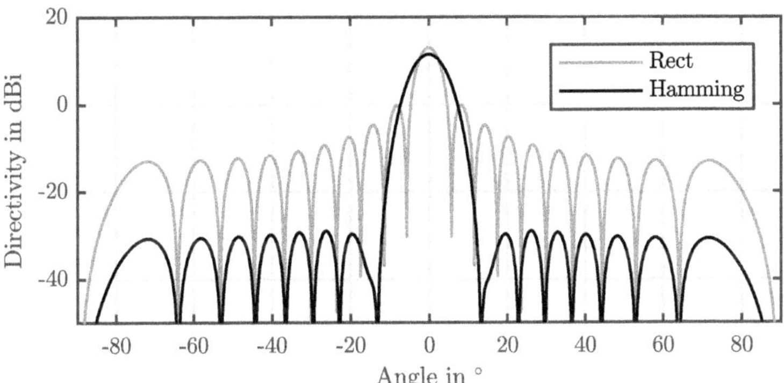

Figure 2.3: Radiation pattern of a fully populated linear array with $\lambda/2$ spacing, without (gray) and with (black) Hamming window

[4]The calculation of the radiation pattern is introduced in Section 2.10.

2.7 Far-Field Condition

When working with certain antenna parameters and calculating, for example, a beamformer matrix, which will be introduced in Section 2.8, the distance of the target or focus has to be known, in general. However, if the point from where the signals are coming, in a receiving case, is far enough away from the antenna, the waves can be approximated as a planar wavefront compared to the antenna's aperture. The distance at which this approximation is valid (within certain tolerances) is called the far-field distance. The assumption of far-field conditions is often useful, as many equations can be simplified this way, such as the beamforming in the following section.

According to [13, pp. 16], the far-field region can be defined as

$$d_{\mathrm{ff}} \geq \frac{2D^2}{\lambda}, \tag{2.25}$$

with D being the (virtual) aperture of the antenna and λ the wavelength. This estimation is based on a maximum allowed phase error on the antenna's aperture.

When operating at high frequencies and with large apertures, far-field conditions are often not satisfied. For example, at 79 GHz, which will be used through this work, an aperture of 20 cm will have a far-field distance of $d_{\mathrm{ff}} \geq 21$ m. This does not necessarily mean that a far-field beamforming result of a near-field target will render completely useless, as will be shown in the following section. Nevertheless, care has to be taken when operating in near-field conditions, and far-field approximations have to be verified.

2.8 Beamforming

Beamforming is a process where transmit signals, receive signals or both are available in the digital domain and manipulated so that the receive and transmit beam is pointing in a specific direction. In general, this is done by delaying the different signals to form a wavefront in the desired direction for transmission or delaying the signals to compensate for the additional delay due to the antenna positions in the case of reception. In other words: A beamformer compensates the difference in the channel caused by different antenna positions in the array for a given target angle. The signals in/from this direction will interfere constructively, while signals in/from another direction should interfere destructively.

With many beams, it is possible to scan the environment and build a map of reflections in dependence of azimuth, elevation, or both. With an additional range information, it is possible to create a three-dimensional representation of the radar's environment. In the case of a MIMO system, both transmit and receive signals are available for a single measurement, and beamforming can be applied to both in post-processing.

According to [3, p. 73], the baseband channel matrix $\boldsymbol{H}_\delta$ between the transmit antenna at $\boldsymbol{a}_t$ and the receive antenna at $\boldsymbol{a}_r$ for the delay δ can be described as

$$\left(\boldsymbol{H}_\delta^{\mathrm{ff}}\right)_{t,r} \propto e^{j\boldsymbol{k}_{\mathrm{tgt}}^{\mathrm{T}} \cdot (\boldsymbol{a}_t + \boldsymbol{a}_r)}, \tag{2.26}$$

with t and r being the index of the transmit and receive antenna, respectively. The wave vector $\boldsymbol{k}_{\mathrm{tgt}}$ is pointing towards the target and $\|\boldsymbol{k}_{\mathrm{tgt}}\| = {}^{2\pi}/_{\lambda}$. This equation applies only to far-field conditions, as only the direction of the wave is taken into account, but not its location or distance. The sum of $\boldsymbol{a}_r$ and $\boldsymbol{a}_t$ represents the position of a virtual antenna element, which will be explained in Section 2.14. In far-field conditions it does not matter to which antenna a shift in position is applied: If the one antenna would, for example, move closer to the target, reducing the overall path length, the same shift applied to the other antenna would have exactly the same effect.

A delay-and-sum beamformer will compensate for the extra delay for each antenna due to its position compared to an arbitrary center of the antenna array and coherently add all signals from the channel matrix $\boldsymbol{H}_\delta^{\text{ff}}$, as described in detail in [14, pp. 23–32]. In the case of a narrowband signal the delay can be approximated by a phase shift and thus every delay δ can be treated independently, as described in [14, p. 34]. Therefore the beamforming matrix has to have the inverse phase compared to (2.26), resulting in

$$\left(\boldsymbol{B}^{\text{ff}}\right)_{t,r} = e^{-j\boldsymbol{k}^{\text{T}}(\boldsymbol{a}_t+\boldsymbol{a}_r)}, \tag{2.27}$$

where $\boldsymbol{k}$ points in the beamforming direction and also satisfies $\|\boldsymbol{k}\| = {}^{2\pi}/_{\lambda}$.

The condition under which a phase shift can be used as an approximation of a true time is given in [14, p. 34] and is

$$B\Delta T_{\text{max}} \ll 1, \tag{2.28}$$

with ΔT_{max} being the maximum delay between any two paths in the system due to different positions of the antennas and B the bandwidth of the transmitted signal.

In the case of near-field beamforming, the distance of the target to be focused on has to be taken into account. The channel model from (2.26) can be reformulated for that purpose. The part $\boldsymbol{k}_{\text{tgt}}^{\text{T}}(\boldsymbol{a}_t + \boldsymbol{a}_r)$ describes the phase offset caused by the antenna position. For a reflecting object at position $\boldsymbol{p}$ the phase from the transmitter to the target and back to the receiver is defined as

$$\phi_{t,r} = k\left(\|\boldsymbol{p} - \boldsymbol{a}_t\| + \|\boldsymbol{a}_r - \boldsymbol{p}\|\right), \tag{2.29}$$

with $k = \|\boldsymbol{k}_{\text{tgt}}\| = \|\boldsymbol{k}\| = {}^{2\pi}/_{\lambda}$, resulting in in the near-field channel definition of

$$\left(\boldsymbol{H}_\delta^{\text{nf}}\right)_{t,r} \propto e^{jk(\|\boldsymbol{p}-\boldsymbol{a}_t\|+\|\boldsymbol{a}_r-\boldsymbol{p}\|)}. \tag{2.30}$$

The beamforming matrix can be constructed equivalently to (2.27), yielding

$$\left(\boldsymbol{B}^{\text{nf}}\right)_{t,r} = e^{-jk(\|\boldsymbol{p}-\boldsymbol{a}_t\|+\|\boldsymbol{a}_r-\boldsymbol{p}\|)}. \tag{2.31}$$

To perform a near-field beamforming scan over a certain range of angles, one can define the target (i.e., focus point) in spherical coordinates:

$$\boldsymbol{p}_{d,\varphi,\theta} = \begin{pmatrix} d\sin(\theta)\cos(\varphi) \\ d\sin(\theta)\sin(\varphi) \\ d\cos(\theta) \end{pmatrix}, \tag{2.32}$$

where d is the distance to the target or focus point. Often the angles are given in azimuth (γ) and elevation (δ), which relate to θ and φ according to

$$\gamma = -\varphi, \tag{2.33}$$

$$\delta = \frac{\pi}{2} - \theta. \tag{2.34}$$

(2.32) can be defined with azimuth and elevation:

$$\boldsymbol{p}_{d,\gamma,\delta} = \begin{pmatrix} d\cos(\delta)\cos(\gamma) \\ -d\cos(\delta)\sin(\gamma) \\ d\sin(\delta) \end{pmatrix}. \tag{2.35}$$

A comparison between far-field and near-field beamforming is shown in Figure 2.4. The transmit and receive arrays are both located on the y-axis, $\{1^5,7^2,6,1,5\}$ for receive and $\{1,3^2,2\}$ for transmit. The virtual aperture (see Section 2.14) is 74 mm, which yields a far-field distance of 2.9 m at the simulation frequency of 79 GHz. The target is located on the boresight axis with a distance of 1 m, which is in the near-field region of the system. Although the target lies in the nearfield, the two beamforming results are very similar, and the negative impact on the far-field beamforming result is minimal.

2.9 Movement Compensation

Both FMCW and CS modulation based systems require the compensation of moving targets if multiple transmitters are used [15,16]. As the transmitters cannot operate simultaneously, they have to be enabled one after another while all receivers are active. Although multiple frequency ramps are transmitted, one for each transmitter, these form a single chirp in terms of an FMCW or CS modulation scheme. Therefore, these chirps are called a chirp group. However, the normal beamforming expects the chirps of a chirp group to be carried out simultaneously. If there are no moving targets, this does not make a difference, but a moving target would have a different phase between the measurements within the chirp group. This phase difference depends on the radial velocity of the target.

In the case of a CS modulation based radar, one of the first steps during signal processing is to calculate the range-Doppler matrix for each transmitter/receiver combination. This step is done before beamforming is applied. Once the range-Doppler matrix is calculated,

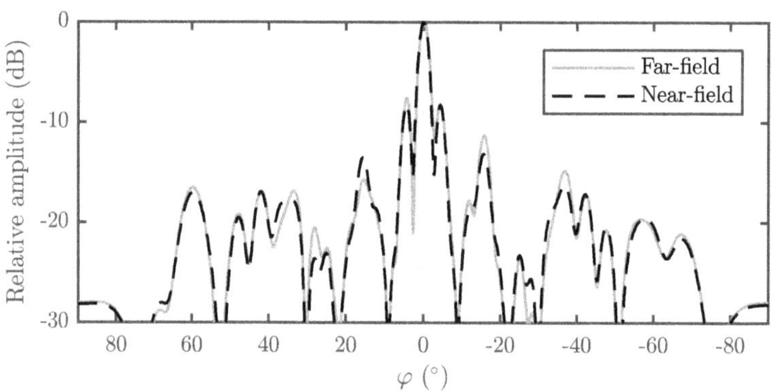

Figure 2.4: Normalized beamforming result for a target in 1 m distance, processed with near-field and far-field beamforming, with the receive array $[1^5,7^2,6,1,5]$ and the transmit array $[1,3^2,2]$, both have $\lambda/2$ spacing at 79 GHz.

the movement compensation can be applied, as the velocity of a target (i.e., for each bin in the matrix) is known. The following discussion is based on a CS radar, but the idea can also be applied to FMCW systems.[5]

As described in Section 2.3 the result of a CS based system is a range-Doppler matrix. This matrix is calculated for each transmitter/receiver combination, usually followed by either target detection or beamforming. The correction is applied to the range-Doppler matrix before any further processing is carried out. The values in the matrices are given as $g_{\delta,f_\mathrm{d},t,r}$, where δ is the distance, and f_d the Doppler. Distance and Doppler are the two axes of the matrices, while t and r are the indices for transmitter and receiver antennas, respectively, building one matrix for each transmitter/receiver combination. The time delay between the first transmitter in the chirp group and the current one is defined as $\Delta\tau_t$.

The Doppler is converted to the radial velocity v_r, according to (2.7). Once the velocity of the target and the transmitter delays are known, these values are converted to the phase value

$$\Delta\Phi_t(f_\mathrm{d}) = 2kv_\mathrm{r}\Delta\tau_t = kf_\mathrm{d}\lambda\Delta\tau_t. \tag{2.36}$$

Consequently, the measured signal is compensated with the known phase offset according to

$$g_{\mathrm{comp},\delta,f_\mathrm{d},t,r} = g_{\delta,f_\mathrm{d},t,r}e^{j\Delta\Phi_t(f_\mathrm{d})}. \tag{2.37}$$

The compensated signal is then used instead of the uncompensated $g_{\delta,f_\mathrm{d},t,r}$ in further signal processing steps.

2.10 Array Factor and Radiation Pattern

The radiation pattern is a property of an array. It describes the radiated signal amplitude, power density, phase, or similar properties as a function of angle. It is possible to express the gain, realized gain, or directivity as a radiation pattern. The radiation pattern might include the process of beamforming, but usually, the radiation pattern is given pointing towards the boresight axis, which corresponds to no beamforming or beamforming towards the boresight axis. As an antenna is usually reciprocal, the same radiation pattern applies to transmission and reception.

The array factor is the influence of the geometry of the array on the radiation pattern. According to (1.50) from [4, p. 14], it is defined as

$$F(\theta,\varphi) = \sum_i w_i e^{jk\boldsymbol{a}_i^\mathrm{T}\boldsymbol{p}_{\theta,\varphi}}, \tag{2.38}$$

with w_i being the weight of the element i, $\boldsymbol{a}_i$ being the position of the element, and

$$\boldsymbol{p}_{\theta,\varphi} = \boldsymbol{p}_{1,\theta,\varphi} = \begin{pmatrix} \sin(\theta)\cos(\varphi) \\ \sin(\theta)\sin(\varphi) \\ \cos(\theta) \end{pmatrix} \tag{2.39}$$

being the vector pointing the in the desired direction with a length of unity. Thus, (2.38) can be extended to

$$F(\theta,\varphi) = \sum_i w_i e^{jk(\boldsymbol{a}_{i,\mathrm{x}}\sin(\theta)\cos(\varphi)+\boldsymbol{a}_{i,\mathrm{y}}\sin(\theta)\sin(\varphi)+\boldsymbol{a}_{i,\mathrm{z}}\sin(\theta))}. \tag{2.40}$$

[5] Due to the long ramps of an FMCW system compared to a CS chirp, the spread of a single target over multiple range-bins has to be taken into account additionally.

Alternatively, the array factor can be written with azimuth and elevation, based on (2.33) and (2.34) as

$$F(\gamma, \delta) = \sum_i w_i e^{jk(\boldsymbol{a}_{i,\mathrm{x}} \cos(\delta)\cos(\gamma) - \boldsymbol{a}_{i,\mathrm{y}} \cos(\delta)\sin(\gamma) + \boldsymbol{a}_{i,\mathrm{z}} \sin(\delta))}. \tag{2.41}$$

The directivity of the array factor is defined as

$$D_{\mathrm{F}}(\theta, \varphi) = \frac{4\pi |F(\theta, \varphi)|^2}{\int_0^{2\pi} \int_0^{\pi} |F(\hat{\theta}, \hat{\varphi})|^2 \sin(\hat{\theta})\, \mathrm{d}\hat{\theta}\mathrm{d}\hat{\varphi}}, \tag{2.42}$$

according to (6-102) in [12, p. 359].

The following calculation will describe the directivity radiation pattern of an antenna array. The radiation pattern is a combination of the properties of the array and the antenna element used within the array. It will be shown that the resulting directivity can be calculated by multiplying the directivity of the element by the directivity of the array using isotropic radiators if certain conditions apply.

According to (1.49) in [4, p. 14], the combined field pattern is calculated by multiplying the field pattern $E_{\mathrm{el}}(r, \theta, \varphi)$ of the element with the array factor:

$$E_{\mathrm{tot}}(r, \theta, \varphi) = E_{\mathrm{el}}(r, \theta, \varphi) F(\theta, \varphi). \tag{2.43}$$

The directivity pattern for any antenna or array given by

$$D(\theta, \varphi) = \frac{4\pi U(\theta, \varphi)}{P_{\mathrm{rad}}}, \tag{2.44}$$

as defined by (2-16) in [12, p. 101], whereas the directivity D is defined as the maximum value of this pattern. P_{rad} is the radiated power of the antenna or array according to (2-13) in [12, p. 38], which is defined as

$$P_{\mathrm{rad}} = \int_0^{2\pi} \int_0^{\pi} U(\theta, \varphi) \sin\theta \; \mathrm{d}\theta \mathrm{d}\varphi. \tag{2.45}$$

$U(\theta, \varphi)$ is the radiation intensity (in Watt per unit solid angle) and is generally defined by (2-12a) in [12, p. 38] as

$$U(\theta, \varphi) = \frac{r^2}{2Z_0} |E(r, \theta, \varphi)|^2, \tag{2.46}$$

with Z_0 being the intrinsic impedance of the medium and $E(r, \theta, \varphi)$ being the far-field electric field strength of the antenna.

Combining (2.43) and (2.46) yields

$$\begin{aligned} U_{\mathrm{tot}}(\theta, \varphi) &= \underbrace{\frac{r^2}{2Z_0} |E_{\mathrm{el}}(r, \theta, \varphi)|^2}_{U_{\mathrm{el}}(\theta, \varphi)} |F(\theta, \varphi)|^2 \\ &= U_{\mathrm{el}}(\theta, \varphi) |F(\theta, \varphi)|^2, \end{aligned} \tag{2.47}$$

describing the radiation intensity of an array with the array factor $F(\theta, \varphi)$ and populated with elements having the radiation intensity pattern $U_{\mathrm{el}}(\theta, \varphi)$.

Inserting (2.47) into (2.44) yields

$$D_{\mathrm{tot}}(\theta,\varphi) = \frac{4\pi U_{\mathrm{el}}(\theta,\varphi)|F(\theta,\varphi)|^2}{P_{\mathrm{rad,tot}}}, \tag{2.48}$$

which is the directivity of an array antenna including the element pattern.

This can also be written as

$$D_{\mathrm{tot}}(\theta,\varphi) = \frac{4\pi U_{\mathrm{el}}(\theta,\varphi)}{P_{\mathrm{rad,el}}}\frac{|F(\theta,\varphi)|^2}{\nu}, \tag{2.49}$$

where $P_{\mathrm{rad,el}}$ is the power supplied to each individual element in the array, not including the weight w_i from the array factor. The weights w_i can be chosen freely, and they influence the radiated power of the array by the array power gain factor ν compared to the actually supplied power. The array power gain is defined as the radiated power of the array using isotropic elements divided by the array's input power:

$$\nu = \frac{P_{\mathrm{rad,array}}}{P_{\mathrm{in,array}}}. \tag{2.50}$$

The radiated power of the array using isotropic elements is given by

$$P_{\mathrm{rad,array}} = \int_0^{2\pi}\int_0^{\pi} U_{\mathrm{iso}}|F(\theta,\varphi)|^2 \sin\theta \; \mathrm{d}\theta\mathrm{d}\varphi, \tag{2.51}$$

according to (2.45). U_{iso} is the constant radiation intensity of an isotropic element. The input power of the array is the same as (2.51) but does not include the array factor:

$$\begin{aligned} P_{\mathrm{in,array}} &= \int_0^{2\pi}\int_0^{\pi} U_{\mathrm{iso}} \sin\theta \; \mathrm{d}\theta\mathrm{d}\varphi & (2.52)\\ &= U_{\mathrm{iso}} 4\pi. & (2.53) \end{aligned}$$

U_{iso} is a constant scalar value in both equations and is independent of θ and φ.

The input power definition in (2.53) of the array is only valid when using isotropic elements, everything is lossless, and the power accepted by the array does not change due to mutual coupling between the elements. The idea behind this approach is the conservation of energy. The power provided to an individual element is now fed into multiple elements and amplified by the array weights due to the arrangement in an array. The factor ν describes the additionally radiated (or fed) power from (or to) the array under the assumption that all elements behave exactly the same within the array as they do stand-alone.

Inserting (2.51) and (2.53) into (2.50) yields the array power gain

$$\begin{aligned} \nu &= \frac{\int_0^{2\pi}\int_0^{\pi} U_{\mathrm{iso}}|F(\theta,\varphi)|^2 \sin\theta \; \mathrm{d}\theta\mathrm{d}\varphi}{U_{\mathrm{iso}}4\pi} \\ &= \frac{\int_0^{2\pi}\int_0^{\pi} |F(\theta,\varphi)|^2 \sin\theta \; \mathrm{d}\theta\mathrm{d}\varphi}{4\pi}, & (2.54) \end{aligned}$$

which is inserted into the directivity definition of the array (2.49):

$$D_{\mathrm{tot}}(\theta,\varphi) = \underbrace{\frac{4\pi U_{\mathrm{el}}(\theta,\varphi)}{P_{\mathrm{rad,el}}}}_{D_{\mathrm{el}}(\theta,\varphi)} \underbrace{\frac{4\pi |F(\theta,\varphi)|^2}{\int_0^{2\pi}\int_0^{\pi} |F(\theta,\varphi)|^2 \sin\theta \; \mathrm{d}\theta \mathrm{d}\varphi}}_{D_{\mathrm{F}}(\theta,\varphi)} . \tag{2.55}$$

Comparing the left part to (2.44) and the right part to (2.42) yields the final definition of pattern multiplication using the directivity patterns:

$$D_{\mathrm{tot}}(\theta,\varphi) = D_{\mathrm{el}}(\theta,\varphi) D_{\mathrm{F}}(\theta,\varphi). \tag{2.56}$$

The gain of an antenna or an antenna array is solely defined as the directivity multiplied with the efficiency η:

$$G(\theta,\varphi) = D(\theta,\varphi)\eta. \tag{2.57}$$

Additionally, the gain of the array is the same as the directivity, as it is lossless. As a consequence (2.56) equivalently applies to the gain of the element:

$$G_{\mathrm{tot}}(\theta,\varphi) = G_{\mathrm{el}}(\theta,\varphi) D_{\mathrm{F}}(\theta,\varphi). \tag{2.58}$$

2.11 Array Factor using FFT

The array factor, especially for two-dimensional problems, is expensive to compute. A more elegant way is to use the discrete Fourier transformation (DFT), implemented with the fast Fourier transformation (FFT) algorithm, to speed up calculations. According to [17, p. 17-5], the DFT is defined as

$$X(u) = \mathfrak{F}_m\{x(m)\} = \sum_{m=0}^{M-1} x(m) e^{-j2\pi m u/M}, \tag{2.59}$$

with $u = 0, 1, 2, ..., M-1$ being the first axis. The two-dimensional DFT can be developed similarly and is defined as

$$X(u,v) = \mathfrak{F}_n\left\{\mathfrak{F}_m\{x(m,n)\}\right\} = \sum_{m=0}^{M-1}\sum_{n=0}^{N-1} x(m,n) e^{-j2\pi m u/M} e^{-j2\pi n v/N}, \tag{2.60}$$

with $v = 0, 1, 2, ..., N-1$ being the second axis. The idea is to compare the array factor in (2.38) to the DFT equation above and match them, so that it is possible to apply a DFT or FFT instead. Again, the array factor is defined as

$$F(\theta,\varphi) = \sum_i w_i e^{jk\boldsymbol{a}_i^{\mathrm{T}} \boldsymbol{p}_{\theta,\varphi}}.$$

First, the array factor is redefined to use an equally spaced two-dimensional grid with the indexes m and n

$$F(\theta,\varphi) = \sum_{m=0}^{M-1}\sum_{n=0}^{N-1} w_{m,n} e^{jk\boldsymbol{a}_{m,n}^{\mathrm{T}} \boldsymbol{p}_{\theta,\varphi}}. \tag{2.61}$$

The element positions are equally spaced on the yz-plane with a distance of $\lambda/2$ according to

$$\boldsymbol{a}_{m,n} = \begin{bmatrix} 0 \\ m\frac{\lambda}{2} \\ n\frac{\lambda}{2} \end{bmatrix}. \tag{2.62}$$

As $\boldsymbol{p}_{\theta,\varphi}$ is a direction vector of unit length, it forms the wave number vector $\boldsymbol{k}$ together with the scalar value k

$$\boldsymbol{k} = k\,\boldsymbol{p}_{\theta,\varphi} = \begin{bmatrix} k_{\mathrm{x}} \\ k_{\mathrm{y}} \\ k_{\mathrm{z}} \end{bmatrix}. \tag{2.63}$$

Inserting (2.62) and (2.63) into (2.61) and solving the scalar product results in

$$F(k_{\mathrm{y}}, k_{\mathrm{z}}) = \sum_{m=0}^{M-1}\sum_{n=0}^{N-1} w_{m,n} e^{jm\frac{\lambda}{2}k_{\mathrm{y}}} e^{jn\frac{\lambda}{2}k_{\mathrm{z}}}. \tag{2.64}$$

When comparing (2.60) with (2.64), the conditions

$$\begin{aligned} x(m,n) &= w_{m,n}, \\ -j2\pi m\frac{u}{M} &= jm\frac{\lambda}{2}k_{\mathrm{y}}, \\ -j2\pi n\frac{v}{N} &= jn\frac{\lambda}{2}k_{\mathrm{z}}, \end{aligned}$$

emerge, which are required for both equations to be equal. Rearranging them results in the three definitions

$$w_{m,n} = x(m,n), \tag{2.65}$$

$$k_{\mathrm{y}} = -\frac{4\pi u}{\lambda M}, \tag{2.66}$$

$$k_{\mathrm{z}} = -\frac{4\pi v}{\lambda N}, \tag{2.67}$$

which, when inserted into the definition of the array factor (2.64), transforms the array factor into the form of a standard two-dimensional DFT. This means that the array factor of a two-dimensional $\lambda/2$-spaced array can be calculated by applying the DFT and converting the DFT axes u and v to the wave number vector with (2.66) and (2.67).

2.12 Side Lobe Level

The SLL is a measure to define the maximum clutter caused by an antenna or an antenna array due to the shape of its beam. It is usually given in dB and is defined as the amplitude difference between the main lobe and the highest side lobe. The side lobe is, by definition, always smaller than the main lobe, resulting in a positive value, given in dB.

In practice, the SLL may be calculated with the radiation pattern or the array factor, depending on the application. The array factor does not include the radiation pattern of the individual antenna elements. If the radiation pattern is used, it can be based on the gain or the directivity. The difference between gain and directivity is just a scalar factor (or offset in dB), which does not influence the SLL, as it is calculated relative to the main lobe's value.

As the SLL is calculated numerically, it is essential to provide sufficient angular resolution for the array factor or radiation pattern not just to resolve every peak but to hit it close to its maximum value. Figure 2.5 illustrates the problem of a too low sample rate in the angular space. To reduce the effect, a higher sample rate in the spatial domain is required. This can be achieved by zero-padding the array structures with additional elements in both directions of zero amplitude. The additional zero elements do not add or change any information in the „input signal" but increasing the number of elements. This results in a higher sample rate in the angular domain. A detailed discussion about zero-padding and its effects can be found, for example, in [18, pp. 90]. This technique can reduce the impact of this problem, but it will never disappear completely. Especially the FFT algorithm is applied when performance is critical and additional zero-padding will reduce the performance. A trade-off between performance and accuracy of the measured peak value needs to be made.

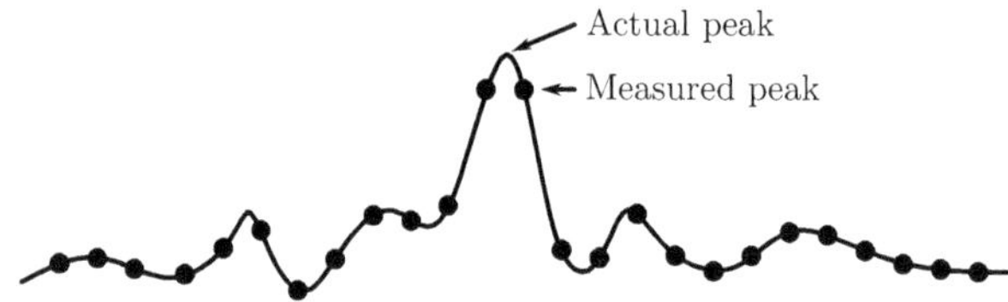

Figure 2.5: Difference between measured peak and actual peak in a discrete system. The line represents the continuous signal, the circles the points of the discrete sampling.

2.13 Synthetic Aperture Radar

To test radar systems with large apertures and a lot of antenna elements, it is advantageous to use a synthetic aperture. Instead of using a physical array, a synthetic aperture is scanned by moving just a single or a few antennas and performing several measurements at positions of the physical array. While a fully populated physical array may perform its measurements simultaneously, the synthetic aperture radar (SAR) carries them out consecutively. These measurements are combined afterward and represent the completely scanned (synthetic) aperture.

The term SAR usually refers to airborne or satellite imaging systems, where the synthetic antenna elements are positioned along a more or less linear path, as described in detail in [19, pp. 17.1] or [20, pp. 835]. This involves specialized and optimized signal processing, which yields a two-dimensional image. One axis is the radar's angular scan (cross-range axis), while the system's range resolution realizes the other axis (range axis).

In general, the principle of the synthetic aperture can be applied to any radar configuration. It is possible to scan a fully populated two-dimensional antenna array with a single transmit antenna and a single receive antenna. These two antennas are placed at all the position combinations of the physical array that is supposed to be represented. Once all combinations are measured, the signal processing is not different from the one with a measurement of the complete physical array. The only significant difference is the incorrect representation of moving targets when the movement is not synchronized to each measurement.

2.14 Mono-static Virtual Aperture

The concept of the virtual aperture of a MIMO radar is well known and described, for example, in [3,21]. This concept's essence is that a MIMO system can be represented solely by a receive array with a single transmitter, or vice versa.

The antennas' positions in the virtual array are calculated by the convolution of the original receive and transmit antenna positions. In a one-dimensional case of the arrays $\boldsymbol{w}_t = [1, 1, 0, 1]$ and $\boldsymbol{w}_r = [1, 0, 1, 1]$, the virtual aperture is $\boldsymbol{w}_v = \boldsymbol{w}_t * \boldsymbol{w}_r = [1, 1, 1, 3, 1, 1, 1]$. This can also be deduced from the definition of the channel in (2.26), which is again

$$\left(\boldsymbol{H}_\delta^{\mathrm{ff}}\right)_{t,r} \propto e^{j\boldsymbol{k}_{\mathrm{tgt}}^{\mathrm{T}}(\boldsymbol{a}_t+\boldsymbol{a}_r)}.$$

The combination of any transmit and receive antenna mimics a virtual antenna at the position

$$\boldsymbol{a}_{\mathrm{bs},t,r} = \boldsymbol{a}_t + \boldsymbol{a}_r. \tag{2.68}$$

It has to be pointed out that this concept is always based on bi-static virtual apertures, meaning that there is only one transmitter at a fixed location, while multiple receivers are forming the virtual aperture or vice versa. It is also possible to calculate a virtual aperture when transmit and receive antenna for a particular combination are at the same position in the virtual array, which may be called the mono-static virtual aperture. A similar discussion can be found in [22]. Therefore the requirement

$$\boldsymbol{a}_t + \boldsymbol{a}_r = \boldsymbol{a}_{\mathrm{t},t,r} + \boldsymbol{a}_{\mathrm{r},t,r}, \tag{2.69}$$

needs to be fulfilled, where $\boldsymbol{a}_{\mathrm{t},t,r}$ and $\boldsymbol{a}_{\mathrm{r},t,r}$ are the antenna positions of the virtual transmitter and receiver, respectively. If the system is a virtual mono-static one, the positions of virtual transmitter and receiver are equal, yielding

$$\boldsymbol{a}_t + \boldsymbol{a}_r = 2\boldsymbol{a}_{\mathrm{ms},t,r} \tag{2.70}$$

for the virtual mono-static position $\boldsymbol{a}_{\mathrm{ms},t,r}$. This results in the final definition of antenna positions in a mono-static virtual aperture

$$\boldsymbol{a}_{\mathrm{ms},t,r} = \frac{\boldsymbol{a}_{\mathrm{bs},t,r}}{2} = \frac{\boldsymbol{a}_t + \boldsymbol{a}_r}{2}. \tag{2.71}$$

This means that the mono-static virtual aperture is the same as the bi-static virtual aperture, with the positions scaled by $1/2$.

A few simulations are carried out to prove the equality between the physical and mono-static virtual aperture. Figure 2.6 shows the beamforming simulation of a single target in a distance of 15 m. The transmit and receive arrays are both minimum redundancy linear arrays (MRLAs), $\{1,3^2,2\}$ and $\{1^5,7^2,6,1,5\}$, respectively. They are located on the y-axis, around the center of the coordinate system. The simulation is carried out at 79 GHz with a bandwidth of 1.5 GHz. It clearly shows that the assumptions on the mono-static virtual aperture are correct. Figure 2.7 shows the same simulation with the difference that the target is only 1 m away from the radar. The bi-static virtual array has an aperture of

74 mm, which results in a far-field distance of 2.9 m, according to (2.25). This simulation shows that (2.71) only holds in far-field conditions. Although the target is as close as one-third of the far-field distance, the results are still very similar and may have limited effect in a practical application.

It also has to be pointed out that the beamforming results for the real aperture shown in Figure 2.6 and Figure 2.7 differ. This is caused by the far-field beamforming algorithm from (2.27), which also influences the result in near-field conditions. However, as only the difference between the real and the mono-static virtual aperture is of interest, this effect can be ignored.

Usually, two separate antennas are used to represent the virtual mono-static element at $\boldsymbol{a}_{\mathrm{ms},t,r}$. They have to be placed next to each other, which may introduce an additional small error. The additional length of the two-way path can be calculated by

$$\Delta d = 2\sqrt{d_{\min}^2 + \left(\frac{D}{2}\right)^2} - 2d_{\min}, \tag{2.72}$$

where D is the distance between the transmit and receive antenna, and $d_{\min}$ is the minimum expected target distance. The effect is negligible, if $\Delta d \ll \lambda$, resulting in the condition

$$D \ll 2\sqrt{\frac{\lambda^2}{4} + \lambda d_{\min}}. \tag{2.73}$$

As $\lambda \ll d_{\min}$ is usually given, the final condition is

$$D \ll 2\sqrt{\lambda d_{\min}}. \tag{2.74}$$

For a distance of 2 m and a wavelength of 3.8 mm the value of the right side of the equation is approximately 17.4 cm. Thus, keeping the distance between the transmitter and the receiver one order of magnitude below this value should have no significant influence on the measurement, and the position of the virtual antenna element can be assumed between the transmitter and receiver.

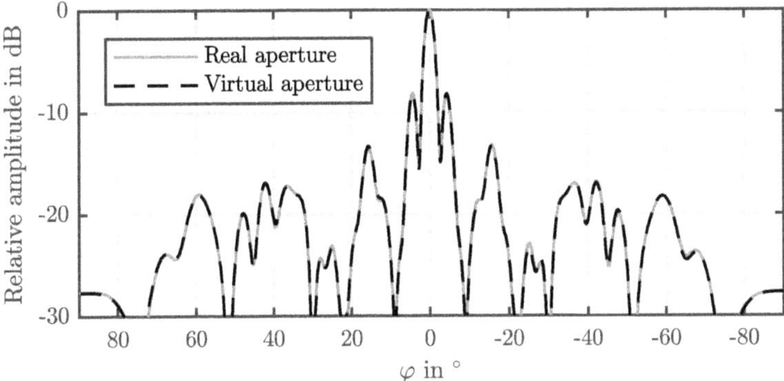

Figure 2.6: Normalized beamforming result for a target in 15 m distance, scanned with either the real or the mono-static virtual aperture with the receive array $[1^5,7^2,6,1,5]$ and the transmit array $[1,3^2,2]$ and $\lambda/2$ spacing at 79 GHz.

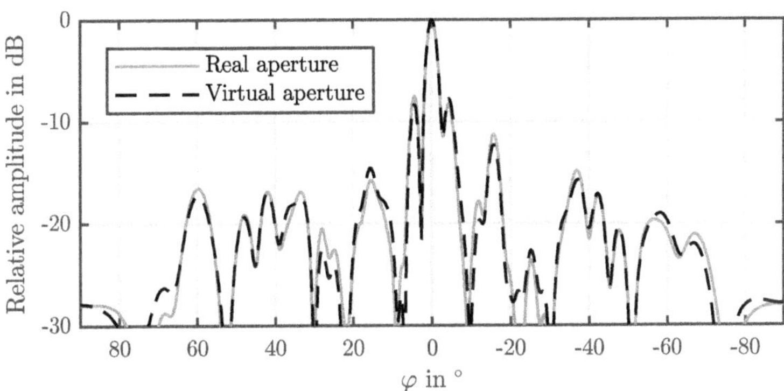

Figure 2.7: Normalized beamforming result for a target in 1 m distance, scanned with either the real or the mono-static virtual aperture with the receive array $[1^5,7^2,6,1,5]$ and the transmit array $[1,3^2,2]$ and $\lambda/2$ spacing at 79 GHz.

2.15 Cell-Averaging CFAR

The CFAR, or more precisely the cell-averaging constant false alarm rate detector, is used to detect targets in a signal with noise present. Every cell is inspected individually, called the cell under test. The noise level P_{n} at this particular point is estimated with the average of the training cells surrounding the cell under test. Additional guard cells between the cell under test and the training cells prevent leakage from the target into the test cells. This would otherwise lead to an overestimated noise level, reducing the sensitivity of the detector.

The following equations are taken from [23, pp. 499]. The noise level is multiplied by a constant factor α to define the threshold P_{T} as

$$P_{\mathrm{T}} = \alpha P_{\mathrm{n}}. \tag{2.75}$$

The noise power is calculated by

$$P_{\mathrm{n}} = \frac{1}{N} \sum_{n=1}^{N} x_n x_n^*, \tag{2.76}$$

where x_n are the sample's (complex) amplitude value of all N_{train} training cells. If the cell under test has more energy than the threshold value, it is considered a target. The algorithm is called constant false alarm rate because it is possible to choose the false alarm rate under the assumption of white Gaussian noise according to

$$\alpha = N_{\mathrm{train}} \left(P_{\mathrm{FA}}^{-1/N_{\mathrm{train}}} - 1 \right), \tag{2.77}$$

where P_{FA} is the probability of a false alarm. The value α is the same as a minimum signal to noise ratio (SNR) as linear power value, thus,

$$\mathrm{SNR}_{\mathrm{min}} = 10 \log(\alpha)\,\mathrm{dB}. \tag{2.78}$$

An example for a two-dimensional cell arrangement is shown in Figure 2.8. It has three guard cells and three training cells in each axis and direction.

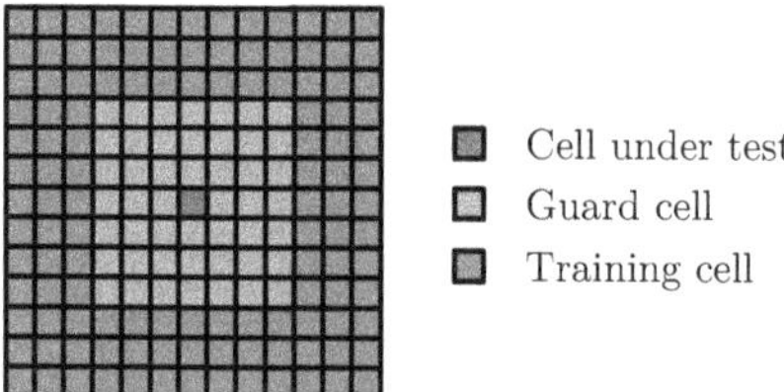

Figure 2.8: Definition of the cell-averaging CFAR cells. The example shows three guard cells and three training cells in both axes.

3 Large Aperture Automotive Radar

The automotive industry is pushing towards highly assisted and even autonomous driving cars [5, 6]. To gather a more precise and reliable representation of the car's surroundings, the sensors themselves and the signal processing are improving over time and are a subject to continuous research. One essential sensor is the radar, which is robust and reliable even in harsh conditions [6]. The primary downside of state-of-the-art automotive radar systems is the low resolution compared to lidar or camera-based systems [24].

To mitigate these drawbacks the resolution of radar systems has to be improved. The bandwidth has to be increased to improve the range resolution (see (2.23)) and the aperture has to be increased to improve the angular resolution (see (2.24)). The primary objective of this chapter is to propose a concept for a radar with enhanced angular resolution. Therefore, the key requirements are defined after introducing the current state-of-the-art, followed by the design of an antenna array, which fulfills the requirements. The array is subsequently characterized, and a signal processing flow for such a system is given.

The following two chapters will implement different aspects of the proposed radar system. Chapter 4 will demonstrate a modular system, the synchronization between the modules, and a reasonably large aperture compared to the state-of-the-art systems. The following Chapter 5 demonstrates the even larger aperture proposed in this chapter, employing a synthetic aperture.

3.1 State of the Art

This section describes the current state-of-the-art for the two fields of interest. Firstly, there is the overview of current radar systems, both in the automotive industry and other areas, which feature similar technology and secondly, the topic of the antenna array's performance and how this array is constructed, called array pattern synthesis.

3.1.1 Radar Systems

Currently the most common bands for automotive radars are located at 24.05–24.25 GHz (200 MHz bandwidth, called the 24 GHz band), 76–77 GHz (1 GHz bandwidth, called the 77 GHz band), and 77–81 GHz (4 GHz bandwidth, called the 79 GHz band) [6]. Using (2.23), this translates to a theoretical range resolution as fine as 37.5 mm for the 79 GHz band. The 79 GHz band is used throughout this work. This value is, of course, not achievable in practical systems. For example, the full bandwidth cannot be used because oscillators have to stabilize before the waveform can be used in FMCW radars. An additionally applied window on the received waveform will also reduce the resolution, as explained in Section 2.6.

Modern radars for adaptive cruise control (ACC) are working in the 77 GHz or 79 GHz band. A typical radar of such a type is the MRRevo14F, manufactured by the Robert

Bosch GmbH. A photo of the antennas and the two frontends is depicted in Figure 3.1. The transmitter on the right is feeding two arrays with a total horizontal aperture of 27 mm. The receiver on the left is fed from four arrays with a horizontal aperture of 16 mm. Thus, the virtual horizontal aperture is 43 mm, if all antennas are used for beamforming. This is equivalent to an HPBW of 4.6 °, according to the estimation given in (2.24). The datasheet [25] claims an object separation capability of 7 ° in azimuth, which is plausible when taking a taper and reduced resolution to the sides into account. This is a typical value for such a radar as stated in [26–29]. Although the MRRevo14F radar module can estimate the elevation angle of an object, the resolution in that axis has to be very limited (or even non-existent), as there is almost no aperture to perform beamforming or beamsteering in the elevation axis. The angle estimation is probably carried out with the small difference in the vertical position of the two transmitter arrays.

The integration of radars into monolithic integrated circuits (ICs) has greatly improved in the past years, and fully integrated ICs have become readily available. There is, for example, the AWR series from Texas Instruments [30, 31], the RXS81 series from Infineon [32] or the TEF810X from NXP [33]. These include everything necessary, like the signal generation, the radar front end, amplifiers, mixers, and ADCs. Some even include signal processing capabilities and are thus capable of reducing the information down to a target list, all within one IC. Therefore, these fully integrated solutions are called radar on chip (RoC). Current research is focused on developing similar chips, but in a higher frequency domain [34–36]. This will be necessary to achieve better resolution in both angular and range while keeping the radar compact. The mentioned papers also show the important change of integrating the antennas directly into the chip, as losses and parasitics in the printed circuit board (PCB) and the PCB interfaces are becoming more problematic at higher frequencies.

A different application using similar technology can be found in the latest full-body scanners for security applications. Such a system has been presented in [37–39]. The final system, which is based on this work, is sold by Rohde und Schwarz GmbH under the names QPS100 and QPS200 [40]. They are both identical, except the first one uses only a single panel of antennas, while the latter one uses two, and the scanned person stands in between them. This way, both sides can be scanned at once. The system operates from 70 GHz to 80 GHz, and each panel consists of 3008 transmit and 3008 receive antennas. It uses a stepped frequency modulation scheme, with one transmitter being active at the same time while all receivers of that panel are receiving. Although this type of full-body scanner has a lot in common with an automotive radar, there are important differences. First off, the measured scenario is completely different, which has a major impact on many aspects of the system. An automotive radar needs much more range but can live with worse angular and range resolution. Typical full-body scanners also do not handle any Doppler information, as this is not required for non-moving targets. More important is the antenna density of the system. The virtual array is more than fully populated; in fact, is it over-populated by a factor of roughly three[1]. This massive spatial oversampling is probably required due to the extreme near-field condition the system is operating in. The situation is not that critical in an automotive radar, which allows the use of sparse arrays. Another difference in this system is the number of produced units and the pricing. The automotive industry has the advantage of potentially requiring millions of such radars each year, while the market for full-body scanners is much smaller. This allows to manufacture relatively new technologies, like the integration of a complete radar system on a chip, at a very low price once mass

[1]Half of one panel is $1\,\mathrm{m}^2$ and has 1504 transmit and 1504 receive antennas. To cover the same area using a simple L-shaped arrangement, 500 antennas would be required on each axis, one for transmitting and one for receiving, using $\lambda/2$ spacing and thus $1504/500 \approx 3$.

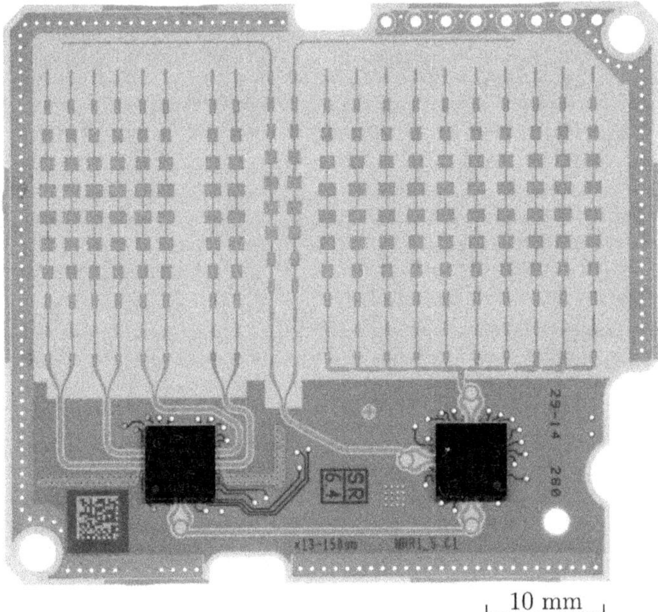

Figure 3.1: Robert Bosch GmbH, MRRevo14F, 77 GHz ACC radar. Receive antennas and frontend (left) and transmit antennas and frontend (right).

production is possible.

3.1.2 Array Pattern Synthesis

This chapter's core topic is the placement of antenna elements in a large two-dimensional array, which suits the particular needs of the intended application. This process is commonly known as antenna pattern synthesis, as a more or less known pattern is given, and the optimal antenna arrangement for the array to fulfill these constraints is wanted. Those constraints are, for example, the beam width or the side lobe level. Array synthesis is a wide field, and solutions to common problems are well known and understood.

Randomly placed array elements are probably the easiest and most straight forward way to achieve a thinned array. There are a couple of different approaches to place the elements randomly. One may define a grid where they can be placed and a probability to place an element at each position in the grid [41, 42]. An example is given in Figure 3.2(a), with a constant probability of 52 %.

A bit more sophisticated is the statistical density taper, as described in [43, pp. 219]. It is based on a continuous taper function instead of a constant value. Window functions like Hamming or Hann are commonly used in such applications. The window's local value is used as a probability for placing an element at a certain position in an equally spaced grid. The array's overall density can be controlled by introducing an additional scaling factor for the window function $0 < k < 1$. This factor controls the probability of an element being placed and, thus, the array's total number of elements. A reduced number of elements will, of course, decrease the array performance, primarily the side-lobe level. An example is shown in Figure 3.2(b) using the Hamming window and a factor of $k = 0.83$.

It is also possible to place the array elements in a non-uniformly manner, also using a window function, which is called the deterministic density taper [43, pp. 212]. In this case, the integral of the window function between two elements is kept constant by varying the distance between them. Figure 3.2(c) illustrates an example, also using the Hamming window, but with $k = 0.93$. The two factors and the probability for the first case are chosen so that all methods yielded an array with 41 antenna elements.

Although only one-dimensional examples are given for these types of more or less randomly placed antennas, applying these methods to a two-dimensional array is possible. The problem with all these methods is that there is no optimization process involved. Thus, it is unlikely that any of these methods will find an optimal solution for a specific problem.

Another type of array that might be considered for such a task is the MRLA, which is not randomly placed but follows strict rules [44]. The elements are placed on a uniform one-dimensional grid. If the array has a length of L, which means there are $L + 1$ potential positions to place an antenna, each distance from 1 to L must be covered at least once by using any pair of populated elements. An MRLA is any array that fulfills this criterion with the minimum possible number of elements for a given length L. This placement method ensures that every distance in the spatial domain is covered using the minimum number of antennas possible. This type of array is known to provide the lowest beamwidth for a given number of antennas [44]. For larger values of L, these arrays become very sparsely populated, e.g. 15.9 % for $L = 112$ [45]. The disadvantage is a relatively high side lobe level, which is plausible given the low density. Additionally, there are only one-dimensional MRLAs, which would require the use of two MRLAs, one for the transmit array and one for the receive array, oriented orthogonally to each other.

There are a few other array synthesis methods to find a (more) optimal solution for a given problem. They usually utilize methods like particle swarm optimization [46–48] or a genetic algorithm [48–50]. Both algorithms seem to be a good approach for such an optimization problem. However, there is little known on such approaches in two-dimensional MIMO arrays. In the current literature, the approach is usually to optimize an individual array, which does not take both the transmitter and the receiver pattern into account, or the algorithms are applied to one-dimensional arrays only. In two-dimensional MIMO arrays, the optimization problem becomes significantly more complex. It has to be noted that none of the above-described methods involves an amplitude weighting, which may help to optimize an array even further.

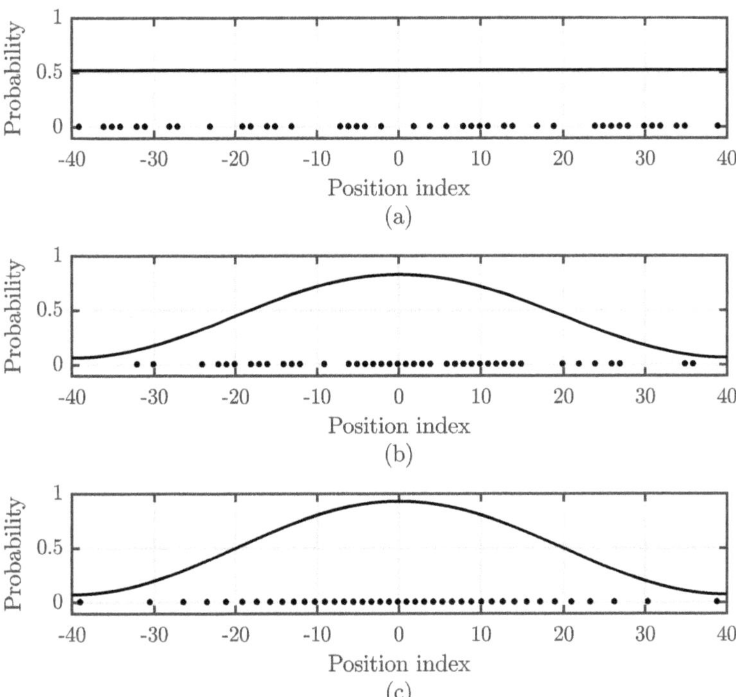

Figure 3.2: Examples for randomly populated array with a probability of 52 % (a), a statistical density taper with a factor of $k = 0.83$ using a Hamming window (b) and deterministic density taper with $k = 0.93$, also using a Hamming window (c). All arrays have 41 antenna elements.

3.2 Requirements

The goal of the proposed radar system is to provide a sufficiently detailed map for autonomous driving. A common distance for such a radar is comparable to radars utilized in ACC applications. An ACC is used to keep the distance to the car in front, even when driving at high speed. A common distance for such applications is 150 m. It would also be beneficial to be able to distinguish between two cars driving on two lanes next to each other. Typically this is done by a separation in range or Doppler, but a resolution of 2.5 m in azimuth would allow an additional dimension of separation. The resolution in elevation is not that critical. However, a certain capability of separating targets in the elevation axis is required to detect, for example, bridges or tunnels.

A resolution of 7 °, like the MRRevo14F presented in the previous section, is equivalent to a width of about 18 m in a distance of 150 m, which is not sufficient for a detailed map of the situation ahead. To achieve the proposed resolution of 2.5 m an angular resolution of 1.0 ° is necessary. This corresponds to an aperture of 19 cm at 79GHz according to (2.24). If even smaller objects should be separable, the aperture has to be increased accordingly. The aperture of 19 cm corresponds to the virtual aperture, and the physical one may be as small as half this value[2]. If the size of this aperture is still too large or has to be made larger to separate closer objects, it is possible to increase the system's operating frequency. For example, at 240 GHz, the aperture would scale down by a factor of three without losing angular resolution.

Due to the availability of hardware and its low cost, the following work is based on 79 GHz components, but almost all of it is also applicable to systems using other frequencies. The aperture is limited to a size, which approximately reaches the 2.5 m azimuth resolution in a distance of 150 m. The resolution in elevation is less critical, but to generate a decent representation of the car's surroundings, it is necessary to keep it in the same order of magnitude.

To build a fully populated virtual array with an aperture of 19 cm at 79GHz, just in a single dimension 126 antenna elements would be required, when using $\lambda/2$ spacing. The number of elements can be reduced by using thinned arrays and a MIMO system. A thinned array has the same aperture as the fully populated one resulting in a similar angular resolution. The drawback of thinning the array is a higher side lobe level. A trade-off between the antenna density (and thus the system complexity and cost) has to be made against the system's dynamic range, which is defined by the side lobes [38].

It is important to note that the side lobes caused by the antenna array of a single target only spread across the azimuth and elevation axes, while the range axis stays relatively clean, even close to the target. The side lobes on the range axis are determined solely by the chosen window function and can be adjusted to almost any requirements during signal processing. Therefore a better range resolution can help to distinguish two targets, which are very close together in both angular and range axis. Once the two targets can be separated in range, the angular side lobes (and also the main lobes) do not interfere with the other target anymore[3].

A MIMO system allows to form a virtual aperture, with an effective number of (virtual) array elements of up to $N_\mathrm{t} \times N_\mathrm{r}$, reducing the number of physical elements down to the

[2] When the transmitter and receiver aperture are each half the required virtual aperture, the required virtual aperture is achieved due to the convolution of the two physical apertures.

[3] The same applies to Doppler. If two targets can be separated by their Doppler value, they can have the same range value without interfering with each other.

square root of the number of virtual elements in the best case. It is also required that all transmit and receive elements are synchronized to allow a coherent measurement between each of them.

3.3 Array Design

As discussed earlier, a couple of different methods are known in the literature for designing large sparse arrays, like a random placement, MRLAs, or some sort of optimizer, which optimizes the placement of the antennas based on some criteria like the side lobe level or the beamwidth. As the optimizer methods are most promising to find an optimal solution, they are the chosen method. In particular, the genetic algorithm is used to optimize both the antenna positions and the weights. Although the genetic algorithm is a widely known concept for solving optimization problems, it leaves much freedom in designing certain aspects of the algorithm. These details in the algorithm's design are the key for solving a specific problem efficiently and with good results.

3.3.1 Array Topology

As described in the previous section, an effective bi-static virtual aperture of 19 cm in azimuth is targeted, while the elevation aperture can be smaller. It is assumed that some sort of fully integrated radar will be used, providing three transmitters and four receivers per chip, which is typical for available solutions like the AWR1243P [31], or the TEF810X [33]. The operation frequency is 79 GHz, yielding a wavelength of 3.8 mm. The chosen modular topology is shown in Figure 3.3. Two chips are packed into a module, providing six transmitters and eight receivers. Each module provides an area at one end to arbitrarily place the antennas from the transceiver chips. This way, such a radar system stays fairly modular. Only the antenna placement has to be redesigned for each module, while the rest of the electronics can stay the same.

The largest of the two exemplarily radar chips is the AWR1243P, with a package size of 10.5 mm x 10.5 mm, or about 15 mm diagonal according to the datasheet [31]. To relax the requirement a bit, a module width of $19\lambda/2$ or 36 mm is chosen, allowing the placement of two chips rotated by 45° next to each other. The effective bi-static virtual aperture should measure approximately 19 cm horizontally. This would require a physical aperture of at least 9.5 cm, which could be achieved with three modules. As will be shown later, the actual aperture has to be a bit larger due to the distribution of antenna elements in the virtual aperture. Therefore, a total number of four modules in the horizontal axis is chosen, resulting in the topology shown in Figure 3.3. A total number of 48 transmitters and 64 receivers is the result, with a total physical azimuth aperture of 14.4 cm.

The placement of the antennas within the complete array is based on a $\lambda/2$ grid. Although a single module has a width of $19\lambda/2$, the outermost grid positions are not used to leave some room between the modules and provide a decent surrounding ground plane for patch antennas.

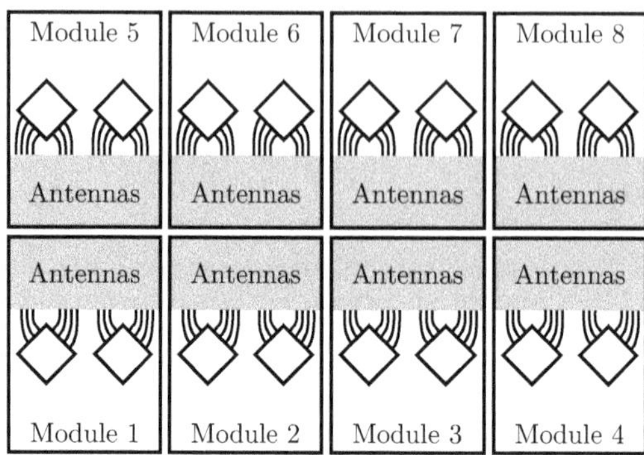

Figure 3.3: Topology of the proposed antenna array

3.3.2 Optimization Algorithm

Now that the radar's general topology and the constraints of antenna placement are known, the optimization process itself needs to be developed. A genetic algorithm is used in this case, as they are known to achieve good results, even in non-linear systems and with local minima being present [51, 52]. The computation-intensive algorithm described below is written in a standalone C++ program.

A genetic algorithm works with populations. A population is, in this case, a set of array configurations, called chromosomes. Each chromosome is holding the positions (or weights) for each of the 112 antennas. A chromosome has a scalar value, called fitness, describing how valuable it is. In general, the fitness function can be defined freely. In this case, the SLL is a good indicator. However, other factors, like the beam width, can be included in the calculation of the fitness. There is much literature on how to combine different goals into a single scalar value, e.g., in [51, pp. 97] or [52]. One of the most common and simplest ways is to add the different values together, using a scaling factor for each one. The tricky part is to determine the scaling factors. Incorrectly chosen values may cause the genetic algorithm to optimize in the wrong direction. A commonly known problem is that the genetic algorithm optimizes for a single variable and ignores the others [52].

As it turned out during testing of the implemented algorithm, it is beneficial to split the optimization process into two stages. The first one takes care of the element placement, while the second one optimizes the antenna weights. It is possible to optimize both parameters in a single optimizer; however, the solution space becomes incredibly large, making the genetic algorithm require a huge population, and the convergence is much slower.

The first population in a genetic algorithm is initialized randomly and sorted by its fitness. The next step is called cross-over, where two chromosomes, called parents, are selected and mate with each other, producing a child. The parents are selected so that chromosomes with a higher fitness value are more likely to be chosen. The cross-over usually involves the random selection of genes (properties of a chromosome, like an antenna position) from either of the two parents for the child.

After the cross-over step follows the mutation stage. The child is randomly mutated, meaning that certain genes are altered. These steps are repeated until enough children are generated to form a new generation. This new generation may additionally contain other chromosomes. It is common practice to include the last population's absolute best chromosomes, making sure the best ones always survive and no degradation is possible. It is also possible to add new random chromosomes to find new good starting points from where to optimize. In the following, all these steps are explained in detail, both for the position and weight optimization process:

- **Random initialization**
  Every module has a list of six transmitters and eight receivers. In the case of position optimization, a new chromosome is created with the antennas at random positions within that module. It is ensured that no antennas are in the same spot. The weights are all set to unity.
  For weight optimization, the antennas' position is already known, and the weight is initialized to a uniform random value of 1 ± 0.3.
- **Fitness function**
  The fitness function is in both cases based on the maximum side lobe level of the array given in dB. The SLL is calculated with the FFT based array factor, as described in Section 2.11. The FFT uses zero-padding to increase the resolution in the angular domain by a factor of two. This allows a reasonable estimation of the SLL without requiring too much additional computation. Two different additional penalties can be subtracted from that value, each having its own scaling factor. The first penalty is for the "roundness" of the generated virtual array, while the second one is linked to the virtual array's redundancy. Both will be explained in more detail later.
- **Cross-over**
  In both cases, an adjustable number of the best chromosomes of the population is used for reproduction. Two parents are randomly selected from this group of best chromosomes to perform a single cross-over.
  For position optimization, the cross-over is carried out for each module and antenna group (Rx or Tx) individually. A random number of antennas is chosen from one parent, while the rest comes from the other parent. After combining the two parents into one child, the antennas are checked for overlaps. Although antennas are not in the same position within a module, they can be in the same position when taken from two different parents. If antennas at the same position are found, one of them is moved to a new random position within that module. This procedure is repeated while the antenna still overlaps another one.
  In the case of weight optimization, the cross-over is calculated using the mean value between the two parents' weights.
- **Mutation**
  While optimizing the position, the algorithm goes through all antennas and decides with an adjustable probability, called the mutation rate, if the current antenna is to be mutated or not. If the result is true, a new random position next to the current one is generated. If there is no other antenna at that position, the antenna is moved there. Otherwise, another random position, which now can be anywhere in the module, is generated until a free spot is found.
  The mutation in the case of weight optimization is done similarly. If the random process decides that an antenna weight has to be mutated, a normal distributed random number with zero mean value is added to the current weight. The mutation rate and the added value's scale are adjustable parameters of the optimizer.

To better understand the behavior of the genetic algorithms, a few more simple scenarios are optimized first. 16 transmitters and 16 receivers are placed in a single module with no spatial restrictions, and the positions are optimized. A population size of $N_{\mathrm{p,pop}} = 1000$ chromosomes is used, while the best $N_{\mathrm{p,keep}} = 200$ are always kept, and $N_{\mathrm{p,new}} = 200$ are always randomly generated. This leaves 600 chromosomes, which are generated using crossover and mutation. The proportion of chromosomes being used for reproduction is $p_{\mathrm{p,repr}} = 25\,\%$. The mutation probability (probability of a position getting altered) is $p_{\mathrm{p,mut}} = 20\,\%$. The fitness function is, in this case, equivalent to the side lobe level:

$$\mathfrak{f}_{\mathrm{pos}} = \mathrm{SLL}. \tag{3.1}$$

It turns out that about $N_{\mathrm{p,gen}} = 10000$ generations are required to converge to a solution. A summary of the parameters is given in Table 3.1.

Table 3.1: Position optimizer parameters for algorithm tests

Parameter	**Meaning**	**Default value**
$N_{\mathrm{p,pop}}$	Population size	1000
$N_{\mathrm{p,keep}}$	Chromosomes to keep	200
$N_{\mathrm{p,new}}$	Chromosomes to randomly generate	200
$N_{\mathrm{p,gen}}$	Generations (max)	10000
$p_{\mathrm{p,repr}}$	Reproduction ratio	25 %
$p_{\mathrm{p,mut}}$	Mutation probability	20 %

Figure 3.4 shows the results of a typical solution. The solutions are always slightly different in the antennas' placement for each run, but the achieved SLL is very similar. Figure 3.4(a) depicts the physical positions for the transmitters (crosses) and the receivers (circles). Figure 3.4(b) shows the resulting virtual array. The numbers represent the number of virtual elements at that position. Figure 3.4(c) plots the convergence of the SLL, which is also the fitness function in this particular case.

The first thing to notice is the clustering of transmitters and receivers into separate groups. This might be a problem when applied to a large aperture, as the clustering of elements might not be desired. However, this problem will vanish once multiple modules are introduced. As elements are bound to a particular module with a given space where they can be placed, it is impossible to form two large groups of transmitters on the one side and receivers on the other.

The algorithm also tends to generate an oval-shaped real (and virtual) aperture. The radiation pattern of such an aperture also looks oval, as shown in Figure 3.4(d). Usually, it is desired to have a square or round aperture. To achieve this, the before mentioned roundness penalty P_{round} is introduced. It is subtracted from the side lobe level with an adjustable factor k_{round}, yielding the new fitness function

$$\mathfrak{f}_{\mathrm{pos}} = \mathrm{SLL} - k_{\mathrm{round}} P_{\mathrm{round}}. \tag{3.2}$$

The roundness penalty P_{round} is calculated by first calculating the center of the virtual aperture

$$\boldsymbol{c} = \frac{1}{N_{\mathrm{t}} N_{\mathrm{r}}} \sum_{t=1}^{N_{\mathrm{t}}} \sum_{r=1}^{N_{\mathrm{r}}} \boldsymbol{a}_{t,r}. \tag{3.3}$$

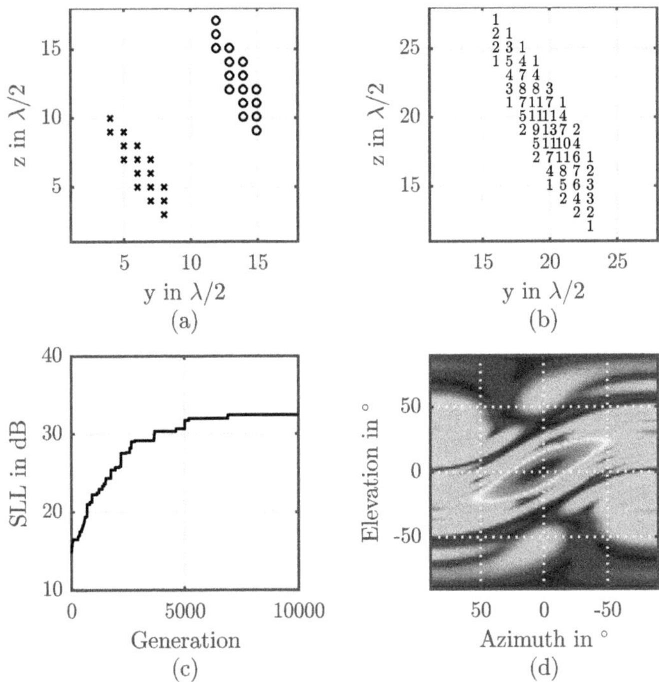

Figure 3.4: Optimized positions of 16 transmitters (crosses) and 16 receivers (circles) using no additional penalty (a), the corresponding virtual aperture (b), the fitness over the number of generations (c), and the radiation pattern of the array with a dynamic range of 60 dB (d)

Next, the virtual antennas are split into four groups $\boldsymbol{a}_{q,n_q}$ corresponding to the four quadrants $q = 1..4$ around the center $\boldsymbol{c}$

$$\boldsymbol{a}_{1,n_1} = \boldsymbol{a}_{t,r} \mid \boldsymbol{a}_{t,r,\mathrm{y}} > \boldsymbol{c}_{\mathrm{y}} \cap \boldsymbol{a}_{t,r,\mathrm{z}} > \boldsymbol{c}_{\mathrm{z}}, \tag{3.4}$$

$$\boldsymbol{a}_{2,n_2} = \boldsymbol{a}_{t,r} \mid \boldsymbol{a}_{t,r,\mathrm{y}} < \boldsymbol{c}_{\mathrm{y}} \cap \boldsymbol{a}_{t,r,\mathrm{z}} > \boldsymbol{c}_{\mathrm{z}}, \tag{3.5}$$

$$\boldsymbol{a}_{3,n_3} = \boldsymbol{a}_{t,r} \mid \boldsymbol{a}_{t,r,\mathrm{y}} < \boldsymbol{c}_{\mathrm{y}} \cap \boldsymbol{a}_{t,r,\mathrm{z}} < \boldsymbol{c}_{\mathrm{z}}, \tag{3.6}$$

$$\boldsymbol{a}_{4,n_4} = \boldsymbol{a}_{t,r} \mid \boldsymbol{a}_{t,r,\mathrm{y}} > \boldsymbol{c}_{\mathrm{y}} \cap \boldsymbol{a}_{t,r,\mathrm{z}} < \boldsymbol{c}_{\mathrm{z}}, \tag{3.7}$$

with N_1, N_2, N_3, and N_4 elements in them, respectively. Next, the center of each of these quadrants relative to the overall center $\boldsymbol{c}$ is defined as

$$\boldsymbol{c}_{\mathrm{q}} = \frac{1}{N_q} \sum_{n=1}^{N_q} \boldsymbol{a}_{q,n_q} - \boldsymbol{c}, \tag{3.8}$$

with q being the quadrant. If any of the absolute axis values $|\boldsymbol{c}_{\mathrm{q,y}}|$ or $|\boldsymbol{c}_{\mathrm{q,z}}|$ is significantly higher or lower than the others, the aperture is somehow deformed. The roundness penalty is therefore defined as the ratio between the maximum and the minimum value of any of these axis values

$$P_{\text{round}} = \frac{\max\{\max\{|\boldsymbol{c}_{q,\text{y}}|\}, \max\{|\boldsymbol{c}_{q,\text{z}}|\}\}}{\min\{\min\{|\boldsymbol{c}_{q,\text{y}}|\}, \min\{|\boldsymbol{c}_{q,\text{z}}|\}\}} - 1. \tag{3.9}$$

The subtracted value of one ensures that the penalty is zero if the aperture is round and increases otherwise. The genetic algorithm now tries to keep the shape of the virtual array fairly round because chromosomes with an oval shape have a lower fitness value and are less likely to reproduce.

The new definition is now applied to the same example as before. The scaling factor $k_{\text{round}} = 1$ is empirically determined to yield good results. The results are shown in Figure 3.5, and the penalty worked as expected. The created antenna positions and the radiation pattern are much less oval. Figure 3.5(c) shows the SLL and the fitness function. As the penalty is subtracted from the SLL, the fitness function is always lower than the SLL. The achieved SLL is slightly lower compared to the one before, as additional restrictions apply on how to position the antenna elements.

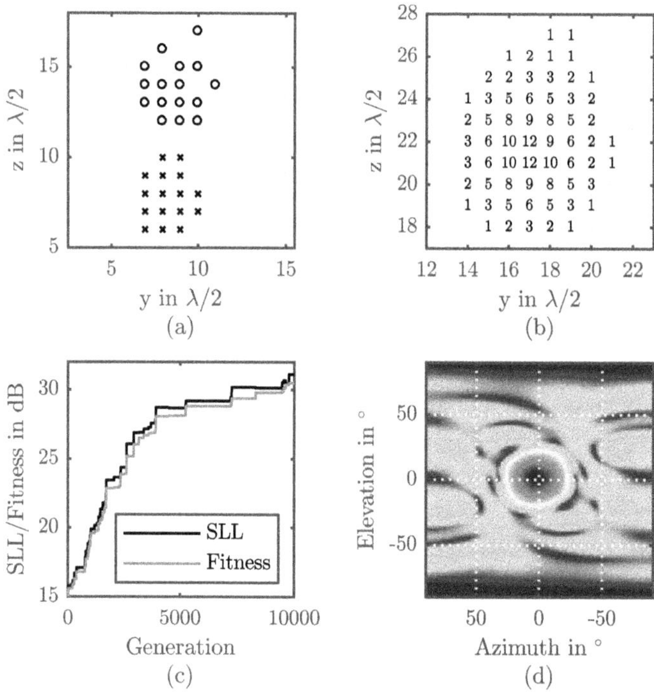

Figure 3.5: Optimized positions of 16 transmitters (crosses) and 16 receivers (circles) using an additional penalty for the roundness(a), the corresponding virtual aperture (b), the SLL and fitness over the number of generations (c), and the radiation pattern of the array with a dynamic range of 60 dB (d)

Another effect observed in both examples is the tendency of the algorithm to form clusters, such that the resulting virtual aperture effectively has a taper function. A lot of virtual elements are located in the center (up to 13 in the first example, see Figure 3.4), but those redundant elements do not add any additional information to the system as the same virtual position is just sampled multiple times. The only purpose they serve is to reduce noise, as this position is measured multiple times, and a beamformer effectively averages these measurements. The taper introduced by this behavior is not required, as it can be replaced by a taper applied as antenna weights during signal processing.

Instead of forming a tapered aperture with multiple antennas and overlapping them, it is more efficient to place the antennas with as little redundancy as possible and apply an artificial taper on the measured samples during signal processing. To force a lower redundancy of the virtual array, a second penalty is introduced to the fitness function, called redundancy penalty P_{red}. It is defined as the ratio between the total number of virtual elements divided by the number of unique virtual elements. Again, the bi-static virtual antenna position is $\boldsymbol{a}_{t,r} = \boldsymbol{a}_t + \boldsymbol{a}_r$, leading to

$$P_{\text{red}} = \frac{\text{count}\{\boldsymbol{a}_{t,r}\}}{\text{unique}\{\boldsymbol{a}_{t,r}\}} - 1. \tag{3.10}$$

Like for the roundness penalty, 1 is subtracted to ensure a value of zero when there are no element positions used multiple times. In this case, the total number of virtual elements is the same as the number of unique elements. This penalty is also subtracted from the SLL with a scaling factor k_{red}, yielding the third and final definition of the fitness function

$$\mathfrak{f}_{\text{pos}} = \text{SLL} - k_{\text{round}} P_{\text{round}} - k_{\text{red}} P_{\text{red}}. \tag{3.11}$$

Applying the new fitness function to the same example as before results in a larger and more equally distributed virtual array. To yield optimal results, the scale factor of $k_{\text{red}} = 3$ is empirically determined, the position mutation probability is lowered to $p_{\text{p,mut}} = 15\,\%$, and the number of generations is reduced to $N_{\text{p,gen}} = 5000$. All other parameters are the same as before. The results are shown in Figure 3.6. As expected, the real and virtual apertures are much larger, and the redundancy is far lower. There are only a couple of positions populated with two virtual antenna elements, compared to the 13 and 12 elements in a single position in the previous examples. With the larger aperture, the HPBW has decreased significantly, as expected.

The downside of this penalty is, of course, the worse SLL compared to the previous runs. The factor k_{red} is a tool to balance between the SLL and the distribution of the antennas (i.e., HPBW). A lower value will yield results with a better SLL and a smaller virtual aperture with more redundant elements.

As mentioned earlier, it is possible to improve the SLL with antenna weights. These weights also increase the HPBW, again leading to a trade-off between both values. To determine the optimal antenna weights, a second genetic algorithm is employed. It uses the fixed positions determined by the position optimizer and only modifies the weights, which were set to 1 during the position optimization. The weights are real-valued only. The use of complex weights did not yield better results, and the imaginary part is usually around zero. The parameters for the weight optimizer are defined in Table 3.2.

Figure 3.7 shows one result of the weight optimizer applied to the last example from Figure 3.6. The weight optimizer is not able to achieve an SLL comparable to the previous results, but the final SLL can be adjusted by using a different value for k_{red} during the

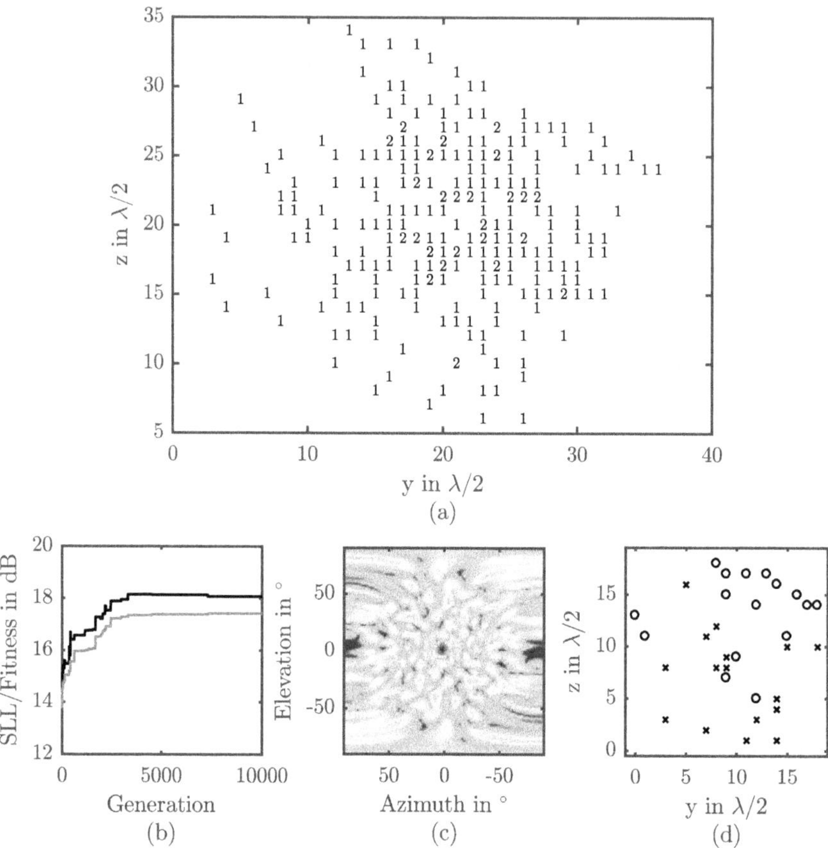

Figure 3.6: Physical (d) and virtual (a) aperture of optimized positions of 16 transmitters and 16 receivers using roundness and redundancy penalty, the fitness over the number of generations (b), and the radiation pattern of the array with a dynamic range of 60 dB (c)

position optimization. In this case, the virtual aperture is very sparse, limiting the side lobe suppression using weights. The SLL increased from 18.1 dB to 22.1 dB during the weight optimization. The HPBW increased slightly from 4.66 ° to 5.06 ° in azimuth, and from 5.36 ° to 5.84 ° in elevation.

It is obviously possible to use both the distribution of the antenna elements and the antenna weights to trade-off between the SLL and the HPBW. The question is: what is the difference between the two methods, and which one is preferred? When assuming a target SLL of, for example, 22 dB, there are two different design methods possible. Either k_{red} is chosen low enough that the position optimizer reaches the goal and no weight optimization is required,

Table 3.2: Weight optimizer parameters

Parameter	Meaning	Default value
$N_{\mathrm{w,pop}}$	Population size	1000
$N_{\mathrm{w,keep}}$	Chromosomes to keep	200
$N_{\mathrm{w,new}}$	Chromosomes to randomly generate	200
$N_{\mathrm{w,gen}}$	Generations (max)	5000
$p_{\mathrm{w,repr}}$	Reproduction ratio	10 %
$p_{\mathrm{w,mut}}$	Mutation probability	20 %
$A_{\mathrm{w,mut}}$	Standard deviation of mutation amplitude	0.05

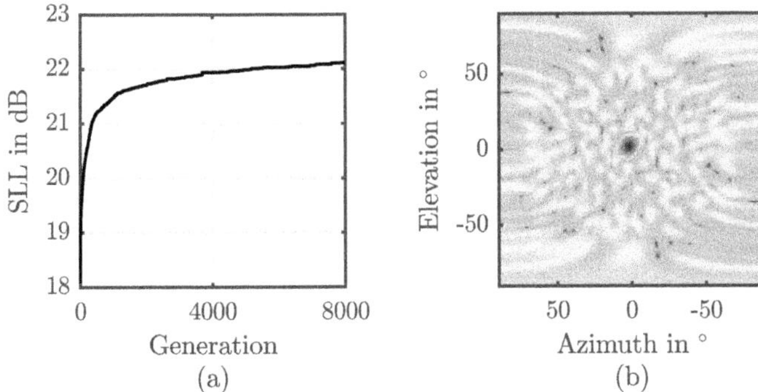

Figure 3.7: Weight optimized results based on the position optimized results from Figure 3.6

or k_{red} is chosen in such a way that the position optimization does not reach the design goal, but the weight optimizer does afterwards.

Figure 3.8 shows the results of both methods with the same scenario as before. The redundancy penalty factor is $k_{\mathrm{red}} = 0.5$ for reaching the SLL without any weight optimization and 3 for using the combination of position and weight optimization. When using only the position optimization, the target SLL of 22 dB is reached within less than 4000 iterations, as seen in Figure 3.8(b). The position optimization of the two-stage optimization process in Figure 3.8(e) reaches a value of 18.3 dB SLL within approximately 5000 iterations. After 10000 generations, the weight optimization is shown in this figure. The short dip directly at generation 10000 is caused by the randomly initialized weights at the beginning of the weight optimization. After that, the genetic algorithm is able to reach the target SLL within approximately 1300 iterations.

When comparing Figure 3.8(a) and 3.8(d) the different approaches clearly influence the way the antenna elements are placed in the array. When applying only the position optimization, the algorithm is forced to group the antennas closely together to form a taper in the virtual array and thus reduce the SLL. The two-stage approach spreads the antenna elements over a larger area.

When comparing the two radiation patterns in Figure 3.8(c) and (f), it is obvious that the

HPBW is far lower when using the combination of position and weight optimization, while the SLL is 22 dB in both cases. In azimuth, for example, the HPBW is reduced from 11.6 ° to 5.9 °, which is a factor of almost two.

In summary, one can say that it is advantageous to increase the redundancy penalty factor k_{red} as far as possible to gain an array with low redundancy and larger aperture. After that, the weight optimizer is applied to increases the SLL as far as possible. If the resulting SLL is not sufficient, k_{red} is reduced, and the process is started again until the desired SLL is reached.

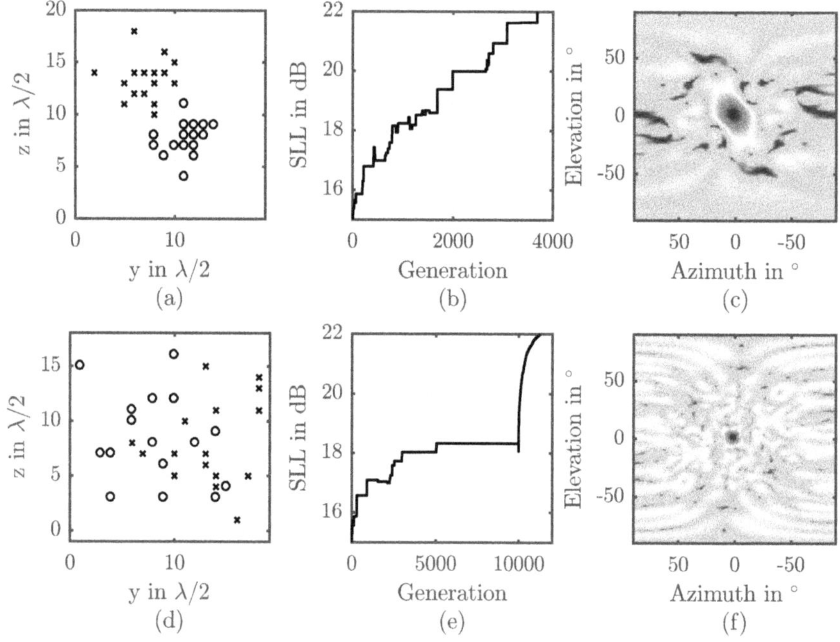

Figure 3.8: Different optimization approaches to reach an SLL of 22 dB. Physical aperture (a), SLL of the position optimization (b), and resulting radiation pattern (c) with no redundancy penalty. Physical aperture (d), SLL of the position and weight optimization combined (e), and resulting radiation pattern (f) with redundancy penalty.

3.3.3 Final Design

The final design for the proposed antenna array uses the modular topology shown in Figure 3.3. It is based on a $\lambda/2$ grid with 4x2 modules. Each module has a width of $19\lambda/2$, while the outermost positions of every module are not used. Table 3.3 shows a summary of the allowed minimum and maximum positions in each axis for each of the modules. Every module holds six transmitters and eight receivers.

Table 3.3: Allowed antenna positions for each module, all coordinates in $\lambda/2$, each as minimum and maximum

Module	y	z	Module	y	z
1	1..18	0..14	5	1..18	16..30
2	20..37	0..14	6	20..37	16..30
3	39..56	0..14	7	39..56	16..30
4	58..75	0..14	8	58..75	16..30

As discussed in the previous section, a combination of position and weight optimization is applied using the two proposed genetic algorithms. The parameters for the genetic algorithms are slightly different to those used earlier, Table 3.4 summarizes all of them. The population size and the number of generations are increased for the position optimization. This is required as the solution space is much larger now. The population needs to be more diverse to cover more potential solutions, and it takes longer to find the optimal one. The reproduction ratio is lowered to achieve a good selection within the larger population. The mutation probability is also lower, as there are more genes present, and changing too many genes at once does not yield good results. The roundness penalty factor is set to zero as the modules provide the overall shape, and the result is expected to be oval, as the aperture is larger in the horizontal axis. The redundancy penalty factor is increased to 10, resulting in good SLL results.

The parameters for the weight optimization are changed similarly. The population size is still 1000, while the number of generations is increased to 50000. The number of newly generated chromosomes for each new generation is set to zero, as it did not improve the algorithm's behavior. The reproduction ratio and mutation probability are both lowered to the same values as for the position optimization. Although many of the changes made to both the parameter sets seem reasonable, most of them are determined empirically.

The results of the optimization process are shown in Figure 3.9. Figure 3.9(a) depicts the physical antenna positions and the module areas. The transmitters are crosses; the receivers are circles. Figure 3.9(b) shows the virtual aperture of the array. The axes for both plots are in mm instead of $\lambda/2$ values as before. The plot for the virtual aperture in Figure 3.9(b) uses gray scales to encode the number of elements in a certain position, as the number would not be readable anymore. There are only black, gray, and light gray, corresponding to three, two, and one element. The physical array is sparsely populated, as expected. The elements are widely scattered, allowing for an easy connection to the individual elements. The Figures 3.9(c) and 3.9(d) show the convergence of the position and weight optimization, respectively.

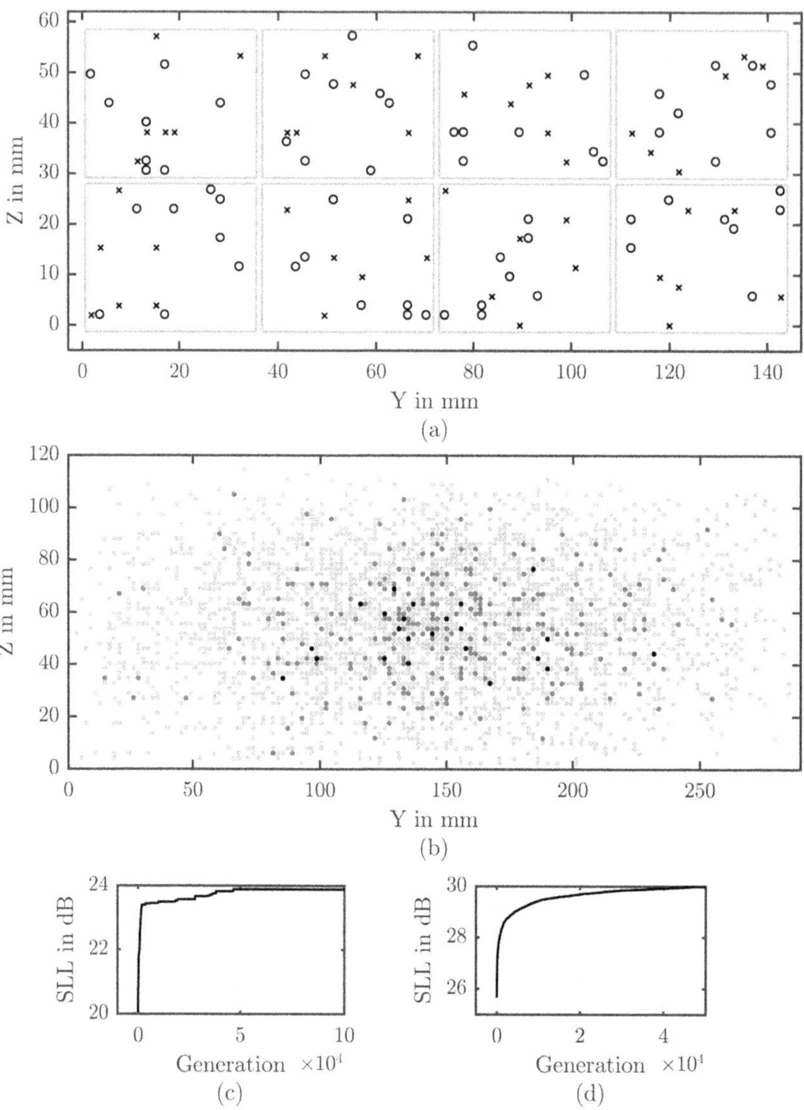

Figure 3.9: Physical transmitter (cross) and receiver (circle) positions (a), virtual array (b), position optimization SLL (c), and weight optimization SLL (d). The brightness of the circles in (b) corresponds to the number of elements at that position. From light gray corresponding to one element up to black corresponding to three elements.

Table 3.4: Optimization parameters for the final design of the antenna array

Parameter	Meaning	Value
$N_{\mathrm{p,pop}}$	Population size	2000
$N_{\mathrm{p,keep}}$	Chromosomes to keep	200
$N_{\mathrm{p,new}}$	Chromosomes to randomly generate	200
$N_{\mathrm{p,gen}}$	Generations (max)	100000
$p_{\mathrm{p,repr}}$	Reproduction ratio	10 %
$p_{\mathrm{p,mut}}$	Mutation probability	5 %
k_{round}	Roundness penalty factor	0
k_{red}	Redundancy penalty factor	10
$N_{\mathrm{w,pop}}$	Population size	1000
$N_{\mathrm{w,keep}}$	Chromosomes to keep	200
$N_{\mathrm{w,new}}$	Chromosomes to randomly generate	0
$N_{\mathrm{w,gen}}$	Generations (max)	50000
$p_{\mathrm{w,repr}}$	Reproduction ratio	10 %
$p_{\mathrm{w,mut}}$	Mutation probability	5 %
$A_{\mathrm{w,mut}}$	Standard deviation of mutation amplitude	0.05

3.4 Array Analysis

The far-field radiation pattern of the proposed antenna is depicted in Figure 3.10. It has a directivity of 34.3 dB, and an HPBW of 0.98 ° in azimuth, and 2.44 ° in elevation. These values fulfill the requirements set in Section 3.2. The depicted far-field radiation pattern corresponds to the virtual aperture, i.e., the combined transmit and receive array. It is impossible to analyze the transmitter and receiver arrays separately, as the weights only apply to the combination of a transmitter and receiver.

However, the virtual aperture sizes from Figure 3.9(b) are 281.8 mm horizontally and 112.4 mm vertically, which translate to HPBW values of 0.69 ° in azimuth and 1.72 ° in elevation, using the estimation from (2.24). The difference can be explained when looking at the distribution of the virtual elements in space. For azimuthal resolution, only the horizontal coordinate (y) of the virtual elements is relevant. Therefore, it is possible to build a linear array by setting the vertical coordinate (z) to zero and adding all the elements' weights in one position. This array has the same radiation pattern in azimuth as the two-dimensional one. This can be proven by setting the elevation δ in (2.41) from page 20 to zero, resulting in

$$F(\gamma) = \sum_i w_i e^{jk(\boldsymbol{a}_{i,\mathrm{x}} \cos(\gamma) - \boldsymbol{a}_{i,\mathrm{y}} \sin(\gamma))}, \tag{3.12}$$

which is independent of the z-coordinate of the elements.

Figure 3.11 shows the vertically added weights spread over the horizontal axis. The shape is very similar to a triangle, which is expected. Both transmitters and receivers are roughly distributed in a rectangular shape. The virtual aperture is the convolution of transmitters and receivers. The convolution of two rectangles is a triangle. According to [10, p. 33], a triangle increases the main lobe width by a factor of 1.27. This leads to an expected angular resolution of 0.87 °, which is much closer to the actual 0.98 °. The remaining difference can

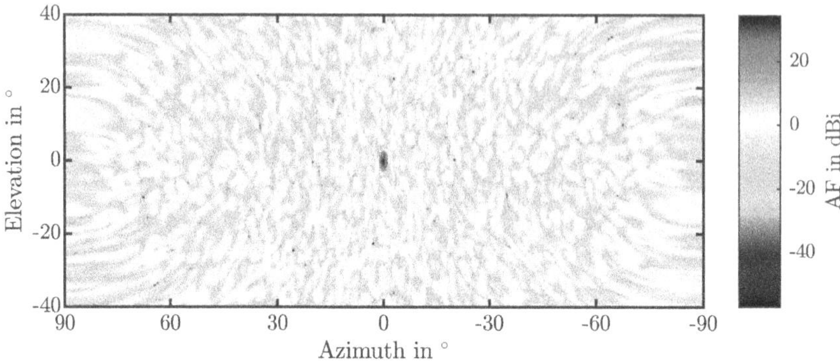

Figure 3.10: Array factor as directivity of the proposed antenna array

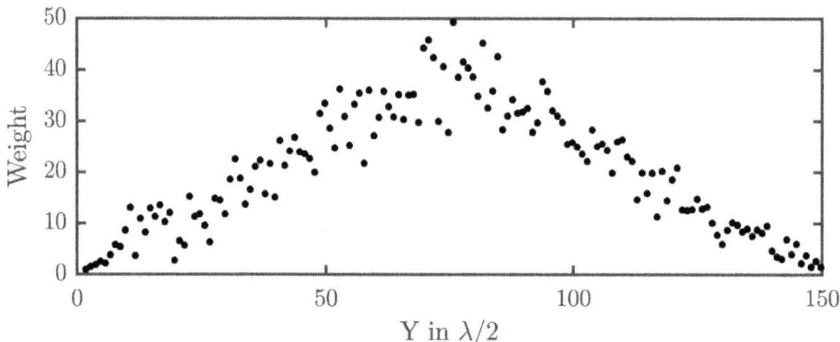

Figure 3.11: Effective antenna weight distribution in azimuth for the proposed antenna array

only be explained by the not perfectly shaped triangle window function, or in other words, the exact weight values of the elements.

The SLL is approximately 29 dB, which is slightly worse compared to the value of 30 dB calculated during the optimization process, shown in Figure 3.9(d). This is caused by the lower resolution the optimizer is using to speed up the calculation, as explained in Section 2.12. The maximum directivity of the array itself (i.e., the directivity when using isotropic antenna elements) is 34.3 dB, which includes transmission and reception. The overall gain will additionally increase by the gains of the transmit and receive antenna elements, for example, the patches.

3.5 Link-Budget

One important measure of the system is the link-budget, or in the case of a radar system, the achievable range. The equation below is a slightly modified radar range equation based on the radar range equation (2.19) on page 13:

$$d < \left[\frac{P_t N_t G_t G_r D_{array} \lambda^2 t_{int} \sigma}{(4\pi)^3 L_{tot} k_0 T_0 F \mathrm{SNR}_{min}} \right]^{1/4} . \tag{3.13}$$

G_t and G_r are the transmit and receive gain of the individual patch antennas, and D_{array} is the directivity (which is also the gain) of the array itself using isotropic elements.

The array's directivity and the number of transmit antennas are known from the previous calculations. All the other parameters depend on the properties of the implemented radar system. Therefore, the parameters from the radar system developed in Chapter 4 are used for the calculation. The radar is using an off-the-shelf AWR1243P RoC from Texas Instruments [31]. A summary of the system's parameters is given in Table 3.5. This includes the transmit power of $P_t = 12\,\mathrm{dBm}$, the gain of the patch antenna elements of $G_t = G_r = 4.5\,\mathrm{dBi}$, additional losses of $L_{tot} = 12.3\,\mathrm{dB}$, and a receiver noise figure of $F = 15\,\mathrm{dB}$. The center frequency is $f_c = 79\,\mathrm{GHz}$, and an SNR of 13 dB is expected.

The last parameter left is the integration time t_{int}, which is harder to estimate as it depends on the application. To give an overview, a few different typical applications are chosen. The first one is equal to the parameters set for both demonstrators, described in Chapters 4 and 5. They utilize an FMCW signal with a duration of $t_s = 51.2\,\mu s$, which is the integration time t_{int} in this case. When using a CS radar one may choose a ramp duration of $t_s = 5\,\mu s$ with 128 chirps, leading to an integration time of $t_{int} = 640\,\mu s$. The ramp duration of 5 µs is based on the assumption that a target may not be faster than 500 km/h, allowing for a maximum ramp periodicity of $T_s = 6.9\,\mu s$, according to (2.15). The value of t_s is chosen slightly lower to have time for oscillator stabilization and wait for the signal to come back from every target before the measurement is started.

Table 3.5: Parameters for the range estimation of the proposed antenna array

Parameter	Meaning	Value
N_t	Number of transmit antennas	48
N_r	Number of receive antennas	64
D_{array}	Transmit and receive array directivity	34.3 dB
G_t, G_r	Transmit/receive element gain	4.5 dBi
f_c	Center frequency	79 GHz
P_t	Transmit power (per transmit antenna)	12 dBm
L_{tot}	Total losses (without antennas)	12.3 dB
F	Noise figure	15 dB
T_0	Temperature	300 K
SNR	Signal to noise ratio	13 dB
t_{int}	Integration time	51.2 µs / 640 µs

All required system parameters are now defined, but the target's reflectivity also influences the achievable range. Table 3.6 lists a few examples of typical targets, their RCS σ, and the expected maximum range d_{max} for both configurations.

Table 3.6: Typical targets, their estimated average cross sections (see sources) and the expected maximum range

Target	σ	d_{max} $t_{int} = 51.2\,\mu s$	d_{max} $t_{int} = 640\,\mu s$
Hammer	−17 dBsm [53]	41 m	76 m
Human/Bicycle	−5 dBsm [8, 54]	81 m	153 m
Bike with pedestrian	−5 dBsm [8]	81 m	153 m
Motorcycle	0 dBsm [8]	108 m	204 m
Car (avg.)	5 dBsm [8]	144 m	271 m
Car (back)	25 dBsm [8]	456 m	858 m

The RCS varies from $-17\,\text{dBsm} = 0.02\,\text{m}^2$ for a hammer as an example of a small metallic object up to $25\,\text{dBsm} = 316\,\text{m}^2$ for the back of a car. Those values are just rough estimates, as they are highly dependent on the angle. The given references for each target define the values more precisely. The predicted ranges for those targets range from 41 m for the hammer up to 858 m for the back of a car. Typical targets on a highway, like a motorcycle or a car, can be detected in a range of at least 153 m, when using the CS modulation signal, which is what the system is designed for.

3.6 Signal Processing Flow

The following signal processing flow is common in typically available radar systems based on CS modulation[4]:

- 2D-FFT on the data matrix for each transmit/receive combination to calculate the range-Doppler matrices.
- Target detection based on individual matrices or the average of the matrices. The average has to be calculated on the absolute values, as the phase would introduce some sort of uncontrolled beamforming.
- Estimation of the AoA for each detected target using beamforming or similar methods.

This signal processing cannot be applied to the proposed system, primarily because the SNR of a single transmitter/receiver combination measurement is not good enough to detect weak targets. The SNR will be increased by the beamforming step significantly as many antennas with relatively low gain are used. In the case of commercially available systems, there are far fewer transmitters, and receivers and the antenna elements in the array have a higher gain. In this case, the difference between the SNR before and after the beamforming step is not as large as in the proposed system.

The proposed system is therefore required to apply beamforming before the target detection is carried out. This requires much more processing power as the beamforming step is one of the more complex ones and needs to be carried out for every point in the range-Doppler matrix instead of applying it only to the points where targets were detected.

[4]Only the important steps are listed. Other things like windowing or zero-padding are usually also applied but are not important for this discussion.

The complete signal processing in case of a CS radar looks like the following:

- Multiplication of a window function over the fast time axis to reduce side lobes in the range axis.
- Zero-padding along the fast and/or slow time axis to increase the angular sampling rate.
- 2D-FFT on the data matrix for each transmit/receive combination to calculate the range-Doppler matrices according to Section 2.3.
- Moving target compensation according to Section 2.9.
- Compensation of amplitude and phase errors based on a calibration, which will be introduced in Section 4.2.
- Multiplication of the antenna weights for each transmitter/receiver combination.
- Beamforming in all necessary directions according to Section 2.8.
- Target detection with the CFAR algorithm according to 2.15.

3.7 Summary

In this chapter, a concept for a large aperture automotive radar was proposed. Section 3.1 described the state-of-the-art in currently available automotive radar systems and the state of current research. Typical automotive radar systems operate in the 77 GHz or 79 GHz band and provide an angular resolution of approximately 7 ° [26–29]. This section also discussed several antenna placement methods in a sparse array with methods like random placement, tapered placement, and optimization. The next section, 3.2, defined the requirements for the array of future automotive radar systems. It has to be a two-dimensional array to generate a three-dimensional representation of the car's surroundings. The angular resolution should be approximately 1 ° in azimuth, while the resolution in elevation is not as critical. This should allow the separation of two cars in a distance of up to 150 m. Section 3.3 describes the design of the the proposed array and the utilized methods. It defined a topology of 4x2 modules, which have to be coherently synchronized. Two genetic algorithms were introduced and used to optimize the antenna placement and the weights of the virtual antenna elements. Two optimization parameters allow the position optimization algorithm to control the shape and the density of the array topology. The final design utilizes 48 transmitters and 64 receivers distributed over the 8 modules and has a virtual aperture of 281.8 mm by 112.4 mm, providing a HPBW of 0.98 ° and 2.44 ° in azimuth and elevation, respectively, reaching the requirements of the system. The total directivity of the array itself, without the antenna elements, is 34.3 dB. Finally, the maximum range of a radar system with the proposed array is estimated for a few system configurations and common targets, which show that the system performs as expected.

To test the proposed large aperture radar antenna array, two demonstrators are built in the next two chapters. These demonstrators will implement certain aspects of the radar system, and the goal is to show the capabilities of the proposed modular automotive radar system. The first demonstrator, presented in the following Chapter 4, shows the modular aspect of the system. It focuses primarily on the synchronization of the modules and demonstrates a fairly large array. The second demonstrator, presented in Chapter 5, implements the proposed antenna array employing a synthetic aperture.

4 Synchronization Demonstrator

The first of the two demonstrators built for this work exhibits the basic principle of a large aperture radar consisting of multiple modules that are coherently synchronized. Each module hosts multiple transmitters and receivers and can be arranged in a multitude of configurations due to their modular nature. The radar system presented in this chapter features an L-shaped aperture, using the transmitters in the vertical axis and the receivers in the horizontal axis. Both transmit and receive arrays are sparse linear arrays optimized to achieve a high angular resolution, compared to state-of-the-art systems in automotive applications, as described in Chapter 3.

The demonstrator also shows the possibility of coherently synchronizing multiple radar modules using a radio over fiber system. In the past decades, optically-based systems have been relatively expensive, as they are based on individual and discrete components. The high price is one reason why optical systems are usually not applied in the automotive industry. However, recent research has shown the possibility of co-integrating photonics and electronics into a single chip [55–58]. This technology can drastically reduce the price of such components, as the manufacturing process becomes more streamlined, and mass production is possible. With optical connections between components or modules, the losses are practically independent of the fiber length. Fibers may also be lighter and especially at higher frequencies more flexible compared to coaxial cables or waveguides. This demonstrator is based on discrete, non-integrated, photonic components, and it proves the basic principle to be valid.

First, the hardware and overall concept for the demonstrator are discussed in Section 4.1. This section includes the radar chips chosen, descriptions of how the modules are built and synchronized, the design of the patch antennas and the antenna array. Section 4.2 discusses the required calibration of the system, followed by the introduction of other important system parameters, the software, and signal processing in Section 4.3. Section 4.4 presents the measurements carried out with this system. Finally, the chapter summarizes the work on the demonstrator in the conclusion in Section 4.5. Parts of this system have already been described in [57–60].

4.1 Hardware and Concept

The demonstrator consists of several modules, similar to those proposed earlier. Each module holds two fully integrated RoC radar front ends, providing up to eight receive and six transmit channels per module. The modules are synchronized to form a fully coherent radar using an radio over fiber (RoF) system. The system components, the construction of the modules, their synchronization using an RoF system, and the antenna array are described in this section. A summary of the system parameters, which will be derived subsequently, is shown in Table 4.1. Figure 4.1 depicts a high-level block diagram. The system consists of five modules, while one master module is used to generate the required reference signals, which are distributed to four slave modules. The slave modules carry the actual transmitters

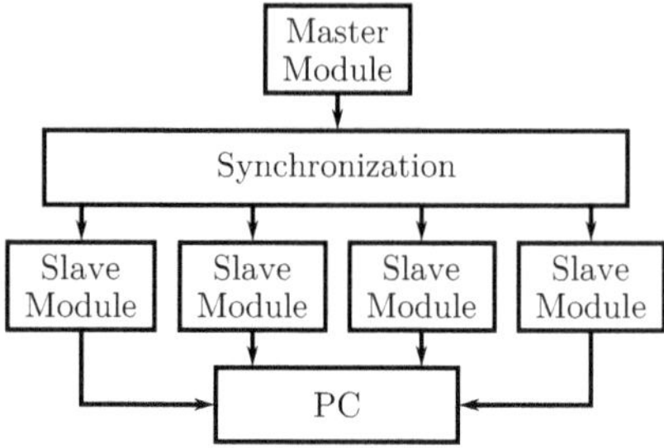

Figure 4.1: High-level block diagram of the synchronization demonstrator

and receivers and are individually connected to a personal computer (PC). The PC receives, stores, and processes the captured radar data. All major components of this demonstrator are listed in the Appendix 8.C.

4.1.1 Radar Chips

The core of the system comprises Texas Instruments AWR1243P fully integrated RoCs. Each chip operates from 76 GHz to 81 GHz and can generate, transmit, and receive an FMCW or CS signal of up to 4 GHz bandwidth. Details about FMCW modulated signals can be found in Section 2.2. After the reception of the transmitted ramp, the signal is amplified, mixed with the reference signal ramp, filtered, and digitized. The digital stream of data is provided either by a camera serial interface (CSI) or a low voltage differential signal (LVDS) interface. Each chip has three transmit and four receive antenna connections. It is supplied in a ball grid array (BGA) flip-chip package, which is directly mounted on a PCB.

The most important feature for this application is the capability of synchronizing multiple chips to a common reference signal, which allows the coherent operation of multiple chips. In single-chip operation, the frequency ramp is generated in the RoC at 1/4 the transmit/receive frequency (approximately 20 GHz), which is then fed into a 4x frequency multiplier before being transmitted or used as reference oscillation for the mixers. This 20 GHz signal can be routed out of the chip and into another one, referred to as master and slave chip, respectively. The slave will use the externally fed reference signal, instead of generating its own. A detailed description of the chip, its features, and its parameters are given in its datasheet [31]. A simplified block diagram showing the vital chip components is shown in Figure 4.2.

The synchronization of the chips requires two signals in addition to the 20 GHz ramp. A common 40 MHz clock is used to generate the internal clocks for components including the synthesizer, which generates the reference ramp signal, the ADC, and other digital logic. The second signal is a trigger to synchronize the start of a measurement, in particular, the simultaneous start of the ADCs.

The AWR1243P is controlled over a serial peripheral interface (SPI), and a portable software framework for its control is provided by Texas Instruments. This framework is written in C and can be integrated into an application running on any system. A few callback functions

Table 4.1: Synchronization demonstrator system parameter summary

Parameter	Meaning	Value	Comment
f_c	Center frequency	78.768 GHz	see 4.3
λ	Wavelength	3.8 mm	see 4.3
s	Ramp slope	29.9813 MHz/µs	see 4.3
t_s	Sweep time	51.2 µs	see 4.3
B	Bandwidth	1.535 GHz	
$f_{s,adc}$	ADC sample rate	5 MHz	see 4.3
N_{ft}	Number of ADC samples	256	see 4.3
B_{bin}	Width of an FFT bin	19.53 kHz	see 4.2
ΔR	Range resolution (no window, calculated)	97.7 mm	see 4.3
ΔR_h	Range resolution (incl. window, calculated)	127 mm	see 4.3
$\Delta R_{h,meas}$	Range resolution (incl. window, measured)	130 mm	see 4.4
$\Delta\gamma_{sim}$	Azimuth resolution (simulated)	1.40 °	see 4.1.4
$\Delta\gamma_{meas}$	Azimuth resolution (measured)	1.37 °	see 4.4
$\Delta\delta_{sim}$	Elevation resolution (simulated)	1.64 °	see 4.1.4
$\Delta\delta_{meas}$	Elevation resolution (measured)	1.67 °	see 4.4
D_r	Receive aperture (horizontal)	133 mm	see Table 4.3
D_t	Transmit aperture (vertical)	110.2 mm	see Table 4.3
D_d	Maximum distance between any Tx/Rx combination	219.9 mm	
d_{ff}	Far-field distance	13.4 m	using D_v
P_t	Transmit power of the RoC	12 dBm	see [31]
F	Noise figure of the RoC	15 dB	see [31]
$l_{max,t}$	Max. microstrip transmitter path length	32.4 mm	see 4.2
$l_{max,r}$	Max. microstrip receiver path length	44.1 mm	see 4.2
L_{us}	Micorstrip losses	0.083 dB/mm	79 GHz
$L_{us,t}$	Max. transmit losses	2.7 dB	see 4.2
$L_{us,r}$	Max. receive losses	3.7 dB	see 4.2
$G_{t,el}$, $G_{r,el}$	Transmit & receive patch gain	4.5 dBi	see 4.1.4

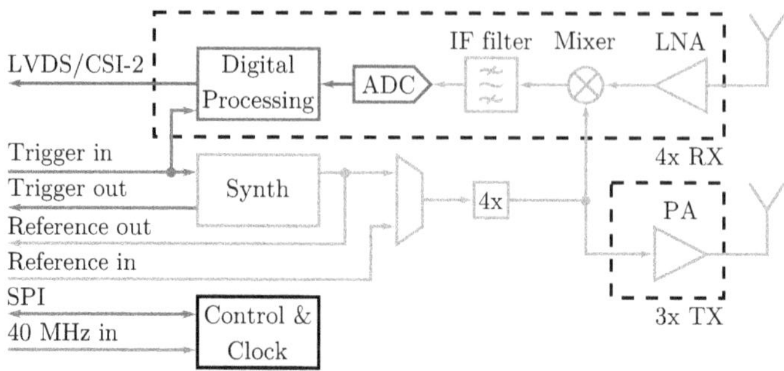

Figure 4.2: AWR1243P simplified block diagram

have to be defined, like the functions for SPI communication or locking features, which are usually provided by an operating system or have to be implemented manually otherwise.

The RoC is capable of performing a fairly complex sequence of chirps. Certain parameters of each chirp can be specified individually, such as the frequency slope, the number of ADC samples to capture and thus the bandwidth of the ramp, the number of repetitions, the ramp start frequency, and many more [31].

4.1.2 Module Hardware

Each module is split into two PCBs mounted on top of each other. A block diagram of the principal module components is shown in Figure 4.3. The bottom PCB is responsible for data capturing, signal processing, and control of the top PCB, which holds the radar front end. The processing PCB is equipped with a Xilinx Artix XC7A50T field programmable gate array (FPGA) to handle the high data flow from the radar chips using the LVDS interface. It buffers and transmits the received data to a PC using a high speed universal serial bus (USB) bridge, the FT232H from FTDI. A DDR3 memory directly connected to the FPGA can be used as an additional buffer for the received waveform data. The board additionally holds a Microchip ATSAM4S2A microcontroller, which is used to control the radar PCB with the framework for the AWR1243P. The top side of the board is shown in Figure 4.5.

The top PCB holds two AWR1243P RoCs as the radar front end. The antennas are patch antennas integrated into the PCB, which will be described in 4.1.4. Both PCBs are additionally equipped with the necessary parts for voltage regulation, interconnects, and debug elements, including light emitting diodes (LEDs) and extra pin headers. During operation, the two boards are stacked onto each other using a high-speed impedance controlled connector. The connector primarily carries the two LVDS and SPI interfaces to the RoCs and a few power rails for various components.

Both PCBs are multi-layer boards. The radar front end PCB has six layers, while the FPGA board has eight layers. Both PCB stacks are described in Figure 4.4 and are made of Isola 185HR substrate, which is similar to standard FR-4 material. For the 79 GHz high-frequency part, Rogers RO3003 material is used for the outer cores of the radar PCB. This material has lower losses and a lower dielectric constant compared to 185HR or standard

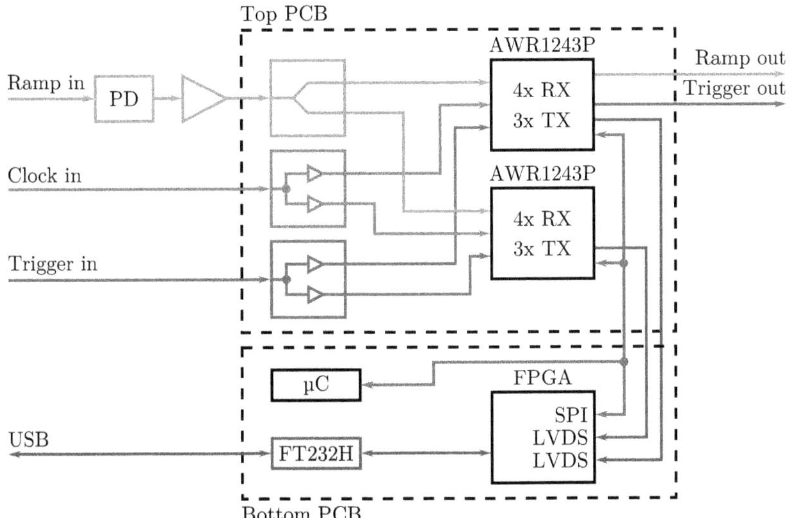

Figure 4.3: Module block diagram of the synchronization demonstrator

FR-4. Figure 4.4 also shows the single-ended 50 Ω and differential 100 Ω line dimensions. The 50 Ω lines are for the antenna feeds, and the 100 Ω differential pairs are used for LVDS and USB. It is worth mentioning that the radar PCB was manufactured with Panasonic R-5775 rather than Isola 185HR due to production issues. This material is slightly different and required a slightly modified stack to achieve the same impedances without changing the layout. This particular stack is shown in Figure 4.4(c).[1]

The two AWR1243P on a single module are already synchronized using the signals described above, thus featuring up to eight receive and six transmit antennas. A block diagram of a single module is shown in Figure 4.3. The two SubMiniature version A (SMA) connectors on the radar board are input (c) and output (e) of the 20 GHz ramp signal, referring to Figure 4.6. The four U.FL connectors are used for the trigger and the common 40 MHz clock signals, each as input and output. In slave operation, only the inputs are used, while a master module generates these signals, depending on its configuration and minor differences in component population. All three synchronization input signals are amplified/buffered before being split and fed to the two RoCs.

The framework for the RoCs runs on the microcontroller located on the FPGA board. The microcontroller is also capable of communicating with the FPGA using an SPI bus. The FPGA will route messages between the PC and the microcontroller along the USB bridge, making it possible to control the RoCs from the PC and get feedback from the microcontroller. The detailed data flow within the modules is described in Section 4.3.

[1] The radar PCB has an additional manufacturing error. The bottom core, designed to be made of RO3003 material, is in fact made of Panasonic R-5775, which primarily influences the 20 GHz reference signal, which is partially routed at the bottom of the PCB. This led to increased losses for the 20 GHz reference signal and required an additional external amplifier.

Cu	Mat.	H	W 50 Ω	W/S 100 Ω
40				
	185HR	127	171	130/170
15				
	185HR	300		
15				
	185HR	115	143	150/100
15				
	185HR	300		
15				
	185HR	124		
15				
	185HR	300		
15				
	185HR	127		
40				

(a)

Cu	Mat.	H	W 50 Ω	W/S 100 Ω
40				
	RO3003	130	276	159/121
15				
	185HR	92		
15				
	185HR	970	131	101/179
15				
	185HR	101		
15				
	RO3003	130		
40				

(b)

Cu	Mat.	H
35		
	RO3003	130
18		
	R-5775	115
18		
	R-5775	1000
18		
	R-5775	115
18		
	RO3003/ R-5775	130
35		

(c)

Figure 4.4: Module PCB stacks for the FPGA PCB (a), the proposed radar PCB (b), and the actual manufactured radar PCB (c). All width (W), height (H), space (S), and copper thickness values are given in µm.

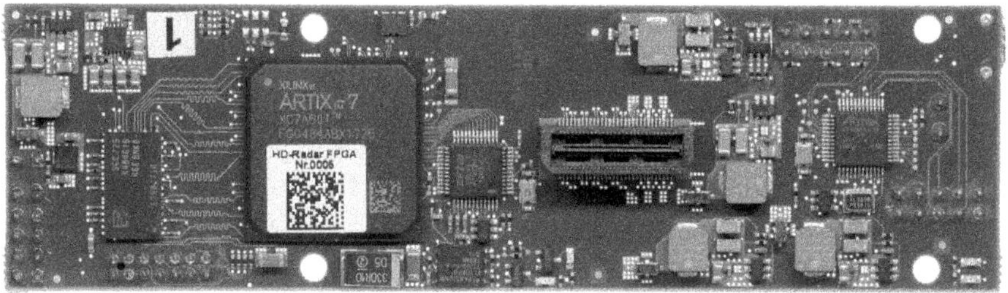

Figure 4.5: Synchronization demonstrator FPGA board holding (from left to right) DDR3 memory, Xilinx Artix-7 FPGA, high speed USB bridge, connector to the top PCB and a Microchip ATSAM4S2A microcontroller

4.1.3 Module Synchronization

As described earlier, each module requires three signals for synchronization, enabling a fully coherent operation of multiple modules. These signals are namely the 20 GHz ramp, the 40 MHz common clock, and the trigger signal. The common clock signal is generated by a central unit using a crystal oscillator and is distributed to all modules. One of these modules, the master, is used to generate the ramp and trigger signal. The trigger is also routed to the central unit, where it is buffered and provided to the slave modules.

Regarding synchronization, the goal of this demonstrator is to show the distribution of a mm-wave radar synchronization signal using an RoF system. The RoF system is implemented for the ramp signal. Because of its high frequency and bandwidth, it is the most complex synchronization signal of all three. The RoF design is described in more detail in [57–60] and will be outlined in the following paragraphs. To reduce the complexity and cost of the demonstrator, the common clock and the trigger signals do not use an RoF system, although possible, and instead incorporate coaxial cables.

Figure 4.7 shows a block diagram of the synchronization elements of the RoF system. The

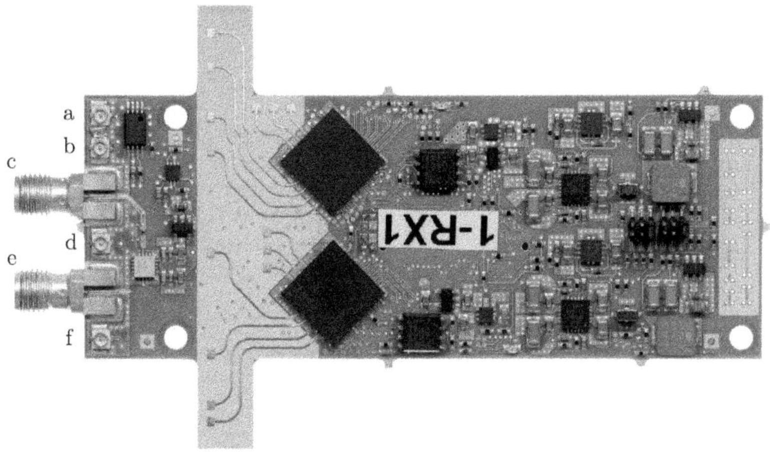

Figure 4.6: One of the synchronization demonstrator's radar boards. The large vertical sparse array is (in this case) the main receive array. The other two arrays with three antennas each are the auxiliary (unused) transmit channels. The connectors on the left are trigger input (a) and output (f), 40 MHz common clock input (b) and output (d), and 20 GHz ramp input (c) and output (e).

core components of the RoF system are the laser diode (LD), the Mach-Zehnder modulator (MZM), the optical amplifier, and the optical splitter. Each slave module is equipped with a photo diode (PD), and an electrical amplifier.[2] The laser diode generates the laser carrier at 1550 nm. The current of the LD is controlled by a laser diode controller (LDC), while the temperature is stabilized with a temperature controller (TEC). The carrier passes through an MZM to modulate it with the ramp signal, which is first amplified. The modulated signal then passes through a 95/5 splitter, and 5 % of the signal is fed into a low-bandwidth PD. This signal is sampled and processed by a microcontroller (µC), which in turn regulates the bias voltage of the MZM to achieve a high degree of modulation. 95 % of the modulated signal is passed through an optical amplifier and split into four paths of equal power. Each of these paths is routed to one of the modules, holding a PD to convert the signal back into the electrical domain. It is amplified and supplied to the slave module. Figure 4.7 additionally shows the distribution of the common 40 MHz reference and trigger signals. The important components and their manufacturers are also listed in Appendix 8.C.

The MZM requires a biasing voltage, which selects the operation point of the modulator. The modulation point has a significant influence on the operation of the modulator. Figure 4.8(a) depicts the measured parameters of the modulator, showing the measured voltage on the monitor PD, which is fed from the 5 % path, and the measured radio frequency (RF) output power of the modulator depending on the bias voltage. A nearly zero bias voltage positions the modulator in the linear region of operation. The result is a fairly linear behavior, while providing a moderate RF output power. As the ramp signal is a swept CW

[2]The additional external amplifier is only required because of the increased loss of the 20 GHz reference signal. This increase in loss is caused by the manufacturer of the radar PCB using the wrong material for the bottom core.

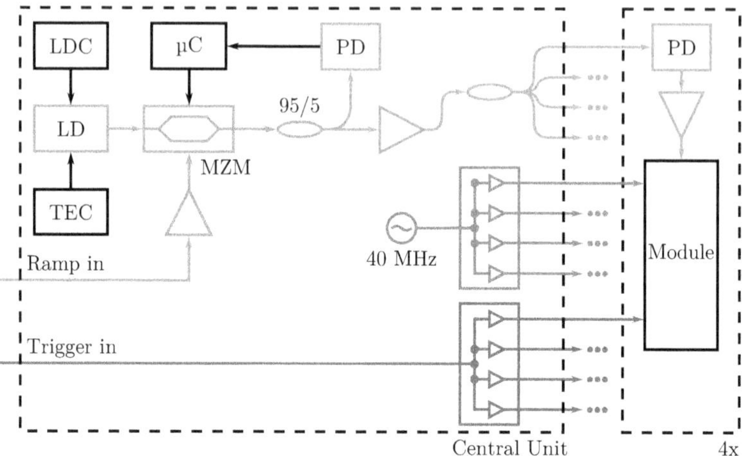

Figure 4.7: Synchronization block diagram, showing the central unit on the left and one out of four modules on the right. The master module, which provides the ramp and trigger signals, is not shown.

tone, it is not susceptible to non-linearities. Potentially generated harmonics are filtered by essentially all of the components in the chain, as the upper operation frequency of most of them does not exceed 20 GHz. It is thus possible to choose a different bias point, providing higher output power. The highest output power occurs close to the minimum monitor voltage and is approximately -9 dBm.

As the operating point of the modulator is highly dependent on the temperature and drifts significantly, the monitor PD is used to build a feedback controller. The relation between the measured monitor voltage and output power is shown in Figure 4.8(b). The target value for the monitor voltage is chosen to be $U_{\mathrm{mon,tgt}} = 10\,\mathrm{mV}$, which is close to the maximum output power, but not too close to the singularity at zero monitor voltage. A simple proportional-integral-differential (PID) controller using only the integral part I solves this problem. The currently set bias voltage is $U_{\mathrm{bias,old}}$, the measured monitor voltage is U_{mon}, resulting in the new bias voltage

$$U_{\mathrm{bias,new}} = U_{\mathrm{bias,old}} + I(U_{\mathrm{mon}} - U_{\mathrm{mon,tgt}}), \tag{4.1}$$

with $I = 0.5$ being determined empirically. A start value of $U_{\mathrm{bias,old}} = -7.5\,\mathrm{V}$ ensures the operating point marked in Figure 4.8.

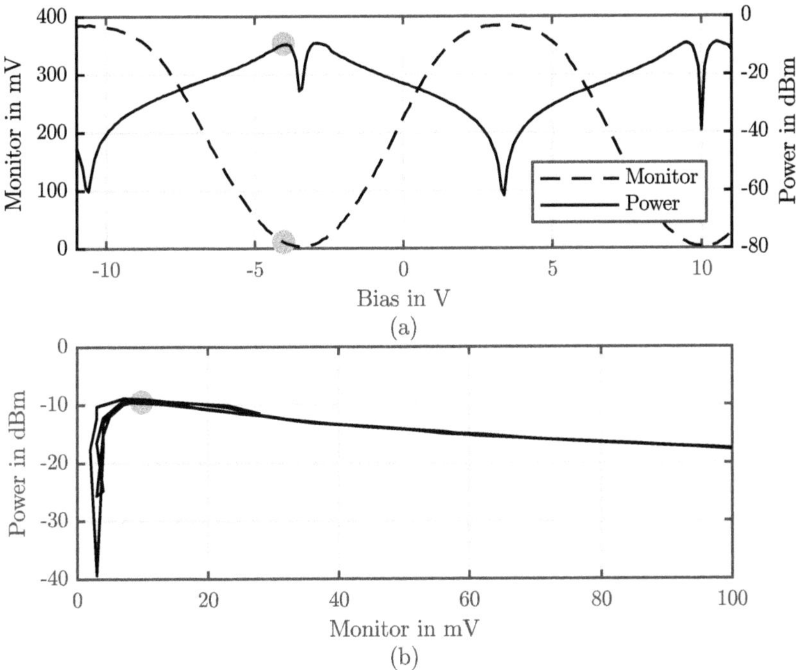

Figure 4.8: MZM monitor voltage and output power in dependence of the biasing voltage (a) and the output power in dependence of the monitor voltage (b). The operation point is marked with a gray circle.

4.1.4 Antenna Elements and Array

The antenna elements are standard PCB patch antennas with slotted feed lines and Rogers RO3003 substrate, the datasheet of which can be found at [61]. The outer core of the radar front end PCB is fabricated from the 5 mil thick variant. Detailed dimensions of the patch antenna are given in Table 4.2, and a corresponding schematic is shown in Figure 4.9. This individual patch has been simulated and optimized using Ansys HFSS. A simulation of the gain, both the E and H plane, is shown in Figure 4.10. The plotted gain is the realized gain and it includes the loss due to mismatch at the antenna's input. The simulated gain in the boresight axis is approximately 4.5 dBi. The patch has adequate input matching in the required band from 76 to 81 GHz of at least -6 dB, as seen in Figure 4.11.

Serially-fed patch arrays are commonly used to increase the gain and narrow the beam, as in the MRRevo14F in Figure 3.1. However, this demonstrator uses only a single patch for each channel for several reasons. One benefit is the higher bandwidth, compared to a serially-fed patch array [62]. Since this demonstrator is used in terms of high-resolution radar, high bandwidth is preferred, improving the range resolution. The lower gain of a single patch can be partially compensated by the number of channels available in the system. Furthermore, single patches are less sensitive to the manufacturing process. As the bandwidth is higher for a single patch, relative errors in the resonance frequency do not significantly affect the frequency operating range. Therefore, a serially-fed patch array usually requires multiple production runs until the array performs as expected, which costs time and money.

The system should have the capability of resolving and distinguishing targets in both azimuth and elevation, as it will be needed for future radar systems. This requires a two-dimensional array with beamforming in both axes. As the antennas cannot be exchanged independently from the modules because they are integrated into the PCB, a separate PCB for each antenna configuration is necessary. As the goal of this demonstrator is not to implement the large array proposed in Chapter 3, simple linear arrays are preferred over the complex two-dimensional arrangement. To achieve two-dimensional coverage using linear arrays, they have to be positioned orthogonally to each other. One array is populated with transmit antennas only, while the other only uses receive antennas. They are arranged in an L shape. As elaborated in Section 2.14, the virtual aperture will form a much more densely populated two-dimensional virtual array by the convolution of transmit and receive elements.

In total, there are six modules built, each with a different antenna layout. Four of these are used for the demonstrator, as the RoF system allows the synchronization of up to four slave modules. A fifth module operates as the master to generate the trigger and ramp signals, while it will not participate in the measurement. Two slave modules form the transmit array, and the other two form the receiver array, resulting in 12 transmit and 16 receive antennas. The relative antenna placement within each array was developed by the project partner perisens GmbH and is described in [60]. The coordinates in multiples of λ for each antenna in its one-dimensional array are given in Table 4.3, and Figure 4.12 shows the full

Table 4.2: Patch antenna dimensions

Symbol	**Name**	**Value**
l_w	Patch width	1300 µm
l_l	Patch length	1060 µm
l_f	Feed width	290 µm
l_s	Slot width	95 µm
l_t	Slot length	340 µm

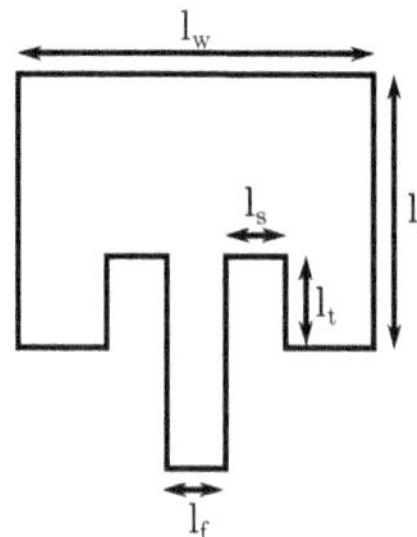

Figure 4.9: Schematic diagram with dimensions of the 79 GHz patch antenna, not to scale

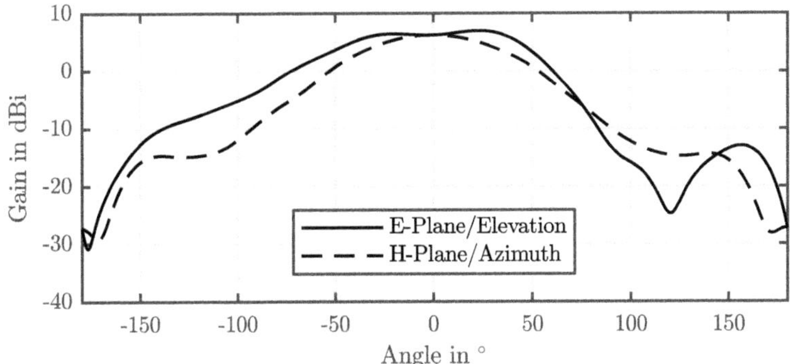

Figure 4.10: Simulated realized total gain of a single 79 GHz patch element in azimuth and elevation

Table 4.3: Patch positions in arrays with mapping to individual modules

Rx Position in λ		**Tx Position in λ**	
Module RX-2	Module RX-3	Module TX-2	Module TX-3
0	15	0	13
1	17	1	15
2	19	2	17
3	21	3	19
4	23	4	21
5	25	8	29
6	27		
10	35		

arrangement of the two-dimensional array in the final coordinate system.

The horizontal array is the receiver, while the vertical array is the transmitter. The simulated far-field radiation patterns for the receive and transmit arrays are plotted in Figure 4.13. Those patterns include the effects of the individual patch antenna, by multiplying the radiation patterns, but no coupling effects between the antennas. The simulated HPBWs are 1.40 ° and 1.64 ° for the receiver (azimuth) and transmitter (elevation), respectively. Both radiation patterns have been primarily designed for a good HPBW, while providing an acceptable side lobe level. Some side lobes further away (roughly ±30 °) are considerably higher, which is common for far-range automotive radar systems as only the center of the radiation pattern is of interest [63].

The combined L-shaped transmit and receive array is depicted in Figure 4.12. The receive array is orientated horizontally, where the antenna with position 0 from Table 4.3 is on the left, and position 35 is on the right from the front view. The transmit array is arranged vertically to the top right of the receive array. Position 0 is at the bottom, and 29 at the top. The distance from the rightmost receive antenna to the lowest transmitter is

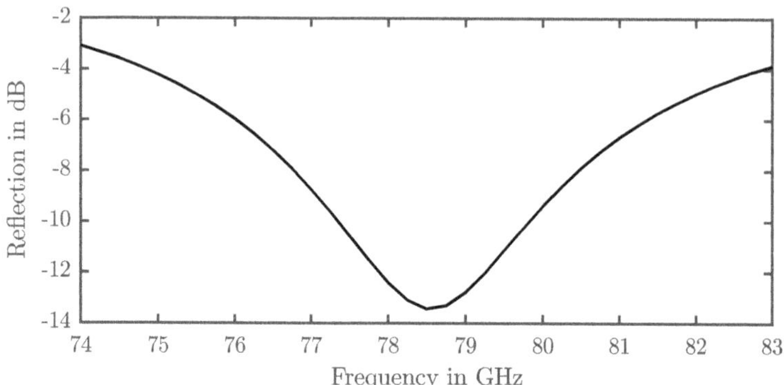

Figure 4.11: Simulated input matching of a single 79 GHz patch element

18.1 mm horizontally and 49.5 mm vertically. Both arrays are polarized vertically, requiring the individual patches on the transmit modules to be rotated by 90 °. Figure 4.14 shows a photo of the demonstrator with the master module on the far left, the two receive modules on the bottom, and the two transmit modules on the right. The overall virtual aperture measures $D_{\mathrm{r}} = 133\,\mathrm{mm}$ horizontally and $D_{\mathrm{t}} = 110.2\,\mathrm{mm}$ vertically. This corresponds to the physical and virtual apertures of the receive and transmit array in this case, as both sub-arrays are linear and arranged in an L shape. The value D_{d} measures 219.9 mm and describes the furthest distance between any transmitter and receiver. In this case it is the distance from the top transmitter to the left receiver, seen in Figure 4.12.

4.2 Calibration

The radar operates at $f_{\mathrm{c}} = 78.768\,\mathrm{GHz}$, which is a wavelength of $\lambda \approx 3.8\,\mathrm{mm}$. For the radar to operate coherently and with correct phase relation between all antennas, the exact wire lengths are needed. Especially the lengths of the optical fibers and the coaxial cables of the synchronization system are not known that precisely. Additionally, other unknown path lengths always exist in the system, such as the electrical lengths of the microstrip lines on the PCBs, which vary within manufacturing tolerances. Therefore, a calibration of the system is required.

There are several ways to implement a calibration for such a system. What they all have in common is that a known reference target is required and the calibration is calculated from the difference between the measurement and a model of this target. A common target is a corner reflector, but others are metal sheets, spheres, or other structures with known reflection properties. The corner reflector has a high RCS that is fairly easy to calculate and is almost independent of the angle of attack, unlike other reflectors. However, the corner reflector is not suitable in this particular case based on the following calculations.

Two conditions are required for a corner reflector to be used as a reference target. First, it needs to be placed in a distance d, such that the reflected signal covers the full receiving aperture D, regardless of which transmitter is active. Otherwise, the receive antennas will

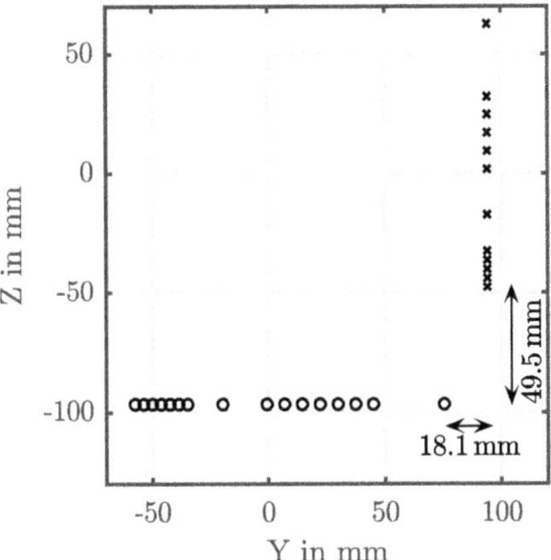

Figure 4.12: Position of the individual patches in the array, front view of the radar system. Crosses are transmitters, circles are receivers.

not be illuminated with the same power density. The HPBW of the corner reflector is a good measure for that. If half of the HPBW covers at least the longest distance between any transmitter and receiver, all antennas receive roughly the same power density. Less than half of the HPBW might be required for this condition, as an antenna right at the edge will only receive half the power density, which might be insufficient for a reliable calibration. The second condition necessary for the corner reflector is that the received power is high enough to be detectable with a sufficient SNR. These two conditions will be examined in detail subsequently and will be combined into a single condition.

The parameters for the HPBW condition are illustrated in Figure 4.15. As explained in Appendix 8.D, the HPBW for a trihedral corner reflector can be estimated with

$$\mathrm{HPBW_{cr}} \approx 66.9\,^{\circ}\frac{\lambda}{a}, \tag{4.2}$$

where a is the size of the corner reflector as defined in Figure 4.15. It is important to note that this equation only holds for far-field conditions, and the following discussion is fully based on far-field approximations.

It is then possible to define a minimum distance for a corner reflector in dependence of its size a and the radar's aperture D, according to Figure 4.15

$$\tan\frac{\mathrm{HPBW}}{2} > \frac{D}{d}, \tag{4.3}$$

thus

$$d > \frac{D}{\tan\frac{\mathrm{HPBW}}{2}} \approx \frac{D}{\tan\left(\frac{66.9\,^{\circ}}{2}\frac{\lambda}{a}\right)} \approx \frac{D}{\tan\left(0.5838\frac{\lambda}{a}\right)}. \tag{4.4}$$

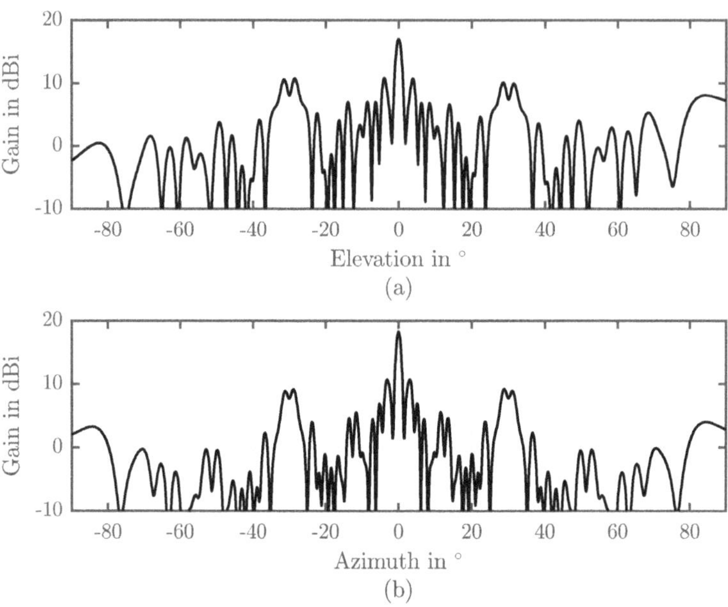

Figure 4.13: Simulated radiation pattern of the transmit (a) and receive (b) array, including the radiation pattern of an individual patch

Figure 4.14: Photograph of the synchronization demonstrator

With $a \gg \lambda$ and $\tan\beta \approx \beta$ for $\beta \ll 1$ follows the first condition:

$$d > 1.7\frac{Da}{\lambda}. \tag{4.5}$$

The second condition requires the corner reflector to be placed close enough such that the received signal is not too weak for proper detection. Especially the calibration signal should be strong compared to noise and clutter. The maximum distance at which a target can be placed, while still being detected, is defined by the radar range equation from (2.16), which is again

$$d < \left[\frac{P_t G_t G_r \lambda^2 \sigma}{(4\pi)^3 P_{r,\min}}\right]^{1/4}. \tag{4.6}$$

including the system losses (see Table 4.1)

$$L_{tot} = L_{us,t} L_{us,r} L_{misc}. \tag{4.7}$$

The losses in the microstrip lines $L_{us,t}$ and $L_{us,r}$ for the transmitters and receivers, respectively, are calculated with the simulation-based loss per mm L_{us} and the maximum lengths $l_{max,t}$ and $l_{max,r}$ of each, resulting in

$$L_{us,t} = L_{us} l_{max,t}, \tag{4.8}$$

$$L_{us,r} = L_{us} l_{max,r}. \tag{4.9}$$

P_t, G_r, and G_t refer to the powers and gains of an individual transmitter/receiver/patch and do not represent the overall system parameters, as the calibration is performed on each transmit/receive combination individually. The minimum receive power $P_{r,\min}$ is defined according to (2.17).

The RCS of a trihedral corner reflector is given in [19, p. 14.10][3] as

$$\sigma = \frac{\pi L^4}{3\lambda^2} = \frac{4\pi a^4}{3\lambda^2}. \tag{4.10}$$

Inserting (4.10) into the radar range equation (4.6), results in

$$d < a\sqrt[4]{\frac{P_t G_r G_t}{3(4\pi)^2 P_{r,\min}}}, \tag{4.11}$$

[3] Defined in the second paragraph of this page. The definition in table 14.1 refers to a dihedral corner reflector.

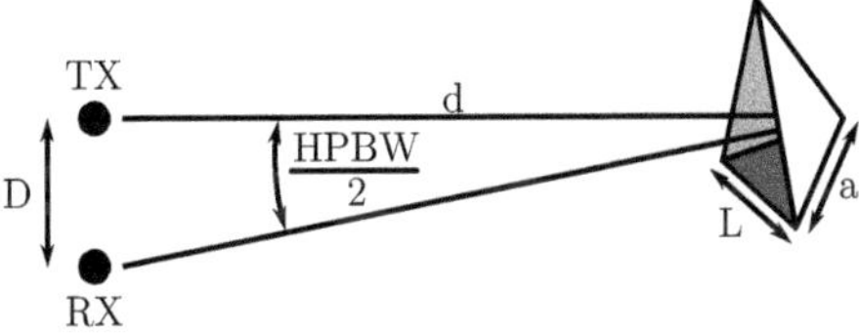

Figure 4.15: Illustration of the main beam of a corner reflector to cover all receivers from any transmitter with half the HPBW or less

which is the second condition the corner reflector needs to satisfy. It is now possible to combine the first condition from (4.5) with the second from (4.11). Rearranging the result yields the final condition for the maximum aperture of the radar as

$$D < \frac{\lambda}{1.7} \sqrt[4]{\frac{P_\mathrm{t} G_\mathrm{r} G_\mathrm{t}}{3(4\pi)^2 P_\mathrm{r,min}}}. \tag{4.12}$$

It is important to note that this condition is independent of the size of the corner reflector and its position. Therefore, only the system parameters determine if a corner reflector can be used as a reference. Most parameters can be taken from Table 4.1. The minimum required SNR is chosen to be 20 dB to achieve good signal quality for calibration, resulting in $P_\mathrm{r,min} \approx -95.9\,\mathrm{dBm}$. The miscellaneous losses L_misc are estimated with 6 dB, yielding $L_\mathrm{tot} \approx 12.3\,\mathrm{dB}$ using (4.7). Inserting the exact values into (4.12) results in

$$D < 198\,\mathrm{mm}, \tag{4.13}$$

describing the maximum aperture the radar is allowed to have. The aperture dimension to compare this value to is D_d from Table 4.1, as it represents the worst-case distance between any transmitter and receiver, which is equal to 219.9 mm and thus slightly larger than the calculated maximum of 198 mm. Therefore, a corner reflector, regardless of its size or what distance it is placed at, cannot be used to calibrate this particular system, as either the received power would be too weak or the corner reflector would focus the beam too much, resulting in different amplitudes for different receive antennas.

Interestingly, the same applies to a flat metal plate. The RCS and the HPBW of a flat metal plate are identical to the corner reflector, except for a scaling factor introduced to compensate for the different definitions of the aperture a, as described in Appendix 8.D. It can be shown that substituting both values in the process described above results in the same condition (4.13), which means that neither a corner reflector nor a metal plate is suitable as a reference target for the calibration process.

However, the resulting requirement in (4.13) only applies to far-field conditions, as many equations used are already far-field approximations, like the RCS and HPBW calculations. There are two potential solutions to this problem: it should be possible to use different, more complex reflectors as reference. Such a reflector should be able to become larger and therefore collect more power, without decreasing the HPBW. A convex-shaped plate might be such a reflector. The other approach, which will be taken here, is to use a reflector in near-field conditions.

Figure 4.16 illustrates the difference between a corner reflector and a plate in near-field conditions using a simple geometrical optics approach. The blue cross is the transmitter, emitting the optic rays. The red circles indicate a traveled distance of 1200 mm for each ray, representing a wavefront. At least in geometrical optics, both targets behave exactly the same if the source is centered relative to the reflector. As things get slightly more complicated for the corner reflector, if the source is not centered and a large plate is easier to construct than a corner reflector, the plate is the preferred target in this case.[4]

The near-field scenario also needs to satisfy the two prior conditions: The reflected signal from any transmitter has to cover all receivers with an amplitude difference as low as possible, and the received power has to be high enough to provide a sufficient SNR.

[4]It has to be pointed out that the metal plate has the disadvantage of an increased dependence on the orientation, which has to be taken into account during the setup of the calibration scenario.

A diagram of the calibration setup is shown in Figure 4.17. It is assumed that the calibration takes place in the near field of the radar. Given are the positions $\boldsymbol{a}_t$ and $\boldsymbol{a}_r$ of the transmitter and receiver, respectively, and the distance d to the reflector. Additionally, it is assumed that all antennas are located in the yz-plane, the radar is directed toward the positive x-axis, the y-axis points to the left when standing behind the radar, and z points upward.

The first condition can still be formulated with (4.3), but the HPBW is defined by the HPBW of the transmitting antenna instead of the corner reflector. The HPBW of the

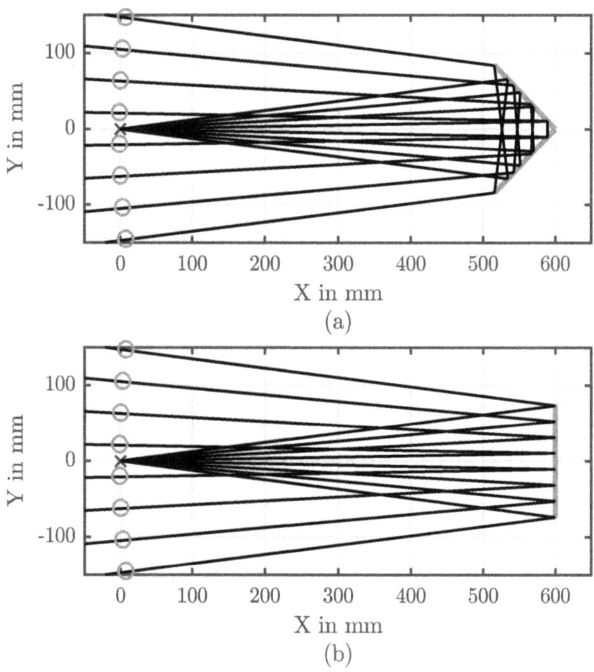

Figure 4.16: Simplified Raytracing of a corner reflector (a) and a plate (b)

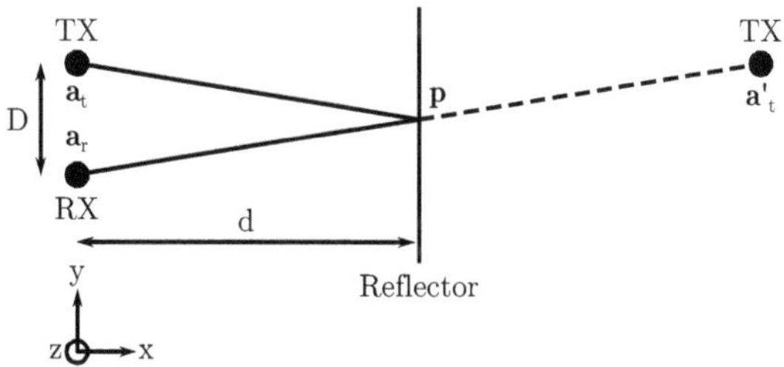

Figure 4.17: Diagram of the calibration scenario in top view including the coordinate system

patch antenna, as depicted in Figure 4.10 on page 63, is approximately 73.6 ° in azimuth, with is the worst-case value compared with the elevation HPBW. Inserting this value and the worst-case aperture $D = D_\mathrm{d}$ into (4.3) yields

$$d > 293.9\,\mathrm{mm}, \tag{4.14}$$

which is the first condition which needs to be satisfied.

For the second condition, the previously used (4.11) cannot be used, as it utilizes the RCS, which is a far-field approximation. Instead, an ideal conducting and infinite plate is assumed, as depicted in Figure 4.17. The assumption of near-field conditions is the key to allow a proper calibration of the system. In that case the RCS can no longer be used as a model to describe the target. The underlying concept of the RCS model is the power density at the target position in combination with the "gain" of the target. The target will collect the power over the RCS cross section and radiate it isotropically. When moving the reflector further away from the transmitter, the collected power decreases and thus the re-radiated power from the target. This behavior is modeled by a $1/R^2$ dependency in the radar equation. Moving the target further away from the radar also decreases the power received by the radar from the target, also with a $1/R^2$ dependency. This leads to the total $1/R^4$ dependency in the radar equation. If a metal plate of infinite size is used as the target, the first $1/R^2$ dependency does not hold any more. The power that hits the plate is always constant and independent of the plate's distance. This reduces the overall distance dependency to $1/R^2$. An other way to look at the effect of an infinite metal plate as reference target is to use image theory [12, p. 182]. In this case the transmitter can be mirrored to the other side of the reflector, and the reflector is removed, resulting in an equivalent scenario. The link between the transmitter and receiver can now be described as a normal radio link with a single path loss and no reflecting target. This removes the dependency on a target's RCS.

The received power is now defined as

$$P_\mathrm{r} = \frac{P_\mathrm{t} G_\mathrm{t} G_\mathrm{r} \lambda^2}{\left(4\pi\sqrt{d^2 + D^2}\right)^2}, \tag{4.15}$$

not including any losses, as they are already taken into account in (2.17), which defines the minimum received power as

$$P_\mathrm{r,min} = k_0 T_0 B_\mathrm{bin} F L_\mathrm{tot} \mathrm{SNR}_\mathrm{min}. \tag{4.16}$$

As the actually received power P_r has to be larger than the minimum receive power $P_\mathrm{r,min}$ (4.15) and (4.16) can be combined to

$$\frac{P_\mathrm{t} G_\mathrm{t} G_\mathrm{r} \lambda^2}{\left(4\pi\sqrt{d^2 + D^2}\right)^2} > P_\mathrm{r,min},$$

yielding

$$d < \sqrt{\frac{P_\mathrm{t} G_\mathrm{t} G_\mathrm{r} \lambda^2}{(4\pi)^2 P_\mathrm{r,min}} - D^2}. \tag{4.17}$$

Inserting the same values as before results in the second condition for a metal plate:

$$d < 51.3\,\mathrm{m}. \tag{4.18}$$

This condition is also easily fulfilled.

The last question is: How large does the plate actually have to be? For a geometrical optics approximation the answer is relatively simple. It has to cover the bi-static virtual aperture of the radar. The position of the virtual antenna element is the same (except for an offset in the x-axis) as the point where the ray will be reflected on the plate. However, it is expected that the reflection of the plate will differ, especially for reflections occurring close to the plate's edges. Figure 4.18 shows the simulation of a 600 mm × 600 mm plate in a distance of 600 mm, as it will be used for the calibration. The simulation is carried out with Ansys Savant, featuring an shooting and bouncing rays (SBR) simulation based on geometrical optics and physical optics. There is a single transmitter located on the boresight axis of the plate. The plot shows the electrical field strength relative to the value at boresight along the y-axis, which is parallel to the plate and in the plane of the radar. The amplitude includes only the rays reflected from the plate and not the source itself. A small ripple on the reflected signal can be observed, which increases to the sides. The ripple will decrease for larger plates. The chosen size of 600 mm seems to be a good compromise, as the ripple is lower than ± 0.6 dB in the relevant area of $\pm D_{\mathrm{d}}$.

Now that a large plate for calibration has been proven a valid reference target, the actual calibration process can be described. To calculate the phase correction term, the measured phase of the target reflection is compared to the prediction. The expected phase can be derived from Figure 4.17 as the path length p the wave has to travel, multiplied by $2\pi/\lambda$. The path p goes from the transmitter at $\boldsymbol{a}_t$ to the reflection point $\boldsymbol{p}$ on the plate and back to the receiver at $\boldsymbol{a}_r$. This is the same distance as the path from $\boldsymbol{a'}_t$ to $\boldsymbol{a}_r$, with

$$\boldsymbol{a'}_t = \begin{pmatrix} 2d \\ a_{t,\mathrm{y}} \\ a_{t,\mathrm{z}} \end{pmatrix}. \tag{4.19}$$

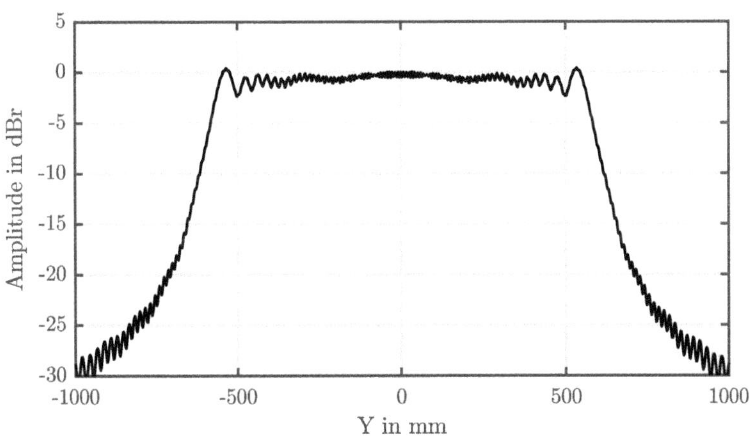

Figure 4.18: Received E-field strength at the radar along the y-axis relative to the boresight axis' value

Thus, the expected phase for a particular transmitter/receiver combination is given by[5]

$$\phi_{t,r,\mathrm{exp}} = \frac{2\pi}{\lambda}\|\boldsymbol{a}_r - \boldsymbol{a'}_t\| = \frac{2\pi}{\lambda}\left\| \begin{pmatrix} 0 \\ a_{r,\mathrm{y}} \\ a_{r,\mathrm{z}} \end{pmatrix} - \begin{pmatrix} 2d \\ a_{t,\mathrm{y}} \\ a_{t,\mathrm{z}} \end{pmatrix} \right\|, \tag{4.20}$$

and the phase offset is the difference between the measured and the expected value:

$$\phi_{t,r,\mathrm{offs}} = \phi_{t,r,\mathrm{meas}} - \phi_{t,r,\mathrm{exp}}. \tag{4.21}$$

The amplitude error is compensated similarly by the ratio of measured to expected value

$$A_{t,r,\mathrm{err}} = \frac{A_{t,r,\mathrm{meas}}}{A_{t,r,\mathrm{exp}}}. \tag{4.22}$$

The expected amplitude value can be assumed constant, as it does not vary significantly for different antenna positions. A complex correction term can be formulated, including the correction for phase and amplitude error

$$C_{t,r} = \frac{1}{A_{t,r,\mathrm{err}}} e^{j\phi_{t,r,\mathrm{offs}}}. \tag{4.23}$$

The calibration is then carried out by multiplying each measured signal sample by the corresponding correction factor $C_{t,r}$.

In general, the purpose of the calibration is to compensate for any unknown delay in the signal path, for example, the different feed lengths to the patch antennas. Phase-shifting is commonly used, as it is easier and faster to compute than true time delays in digital beamformers. The same applies to the calibration process, which is very similar from a mathematical point of view. To use a phase shift instead of a true time delay, (2.28) needs to be fulfilled, which is again

$$B\Delta T_{\mathrm{max}} \ll 1. \tag{4.24}$$

B is the bandwidth of the signal, and ΔT_{max} is the maximum delay difference between any path, including everything from the ramp generation to the transmitter, to the target, back to the receiver and the down conversion.

As the delay can be converted into an equivalent electrical length Δl_{max}, (4.24) can be rewritten as

$$\Delta l_{\mathrm{max}} \ll \frac{c_0}{B}, \tag{4.25}$$

which is approximately 195 mm for the system bandwidth of 1.535 GHz utilized for the measurements in Section 4.4.

Understandably, this equation is very similar to (2.23), the range resolution. (4.25) basically states that a delay has to be significantly smaller than twice the range resolution. Otherwise, a length of twice the range resolution would delay a reflection exactly by one range bin. As all range bins for a particular distance will be added together during beamforming, a single target may fall into multiple bins. This will smear the target in the range axis and thus reduce the range resolution.

[5] The reflection on the wall would cause an additional 180 ° phase shift, which is not taken into account here. This is not required, as only phase differences between the transmitter and receiver are relevant. The additional phase is a constant value, applying to all measured values, and can thus be ignored.

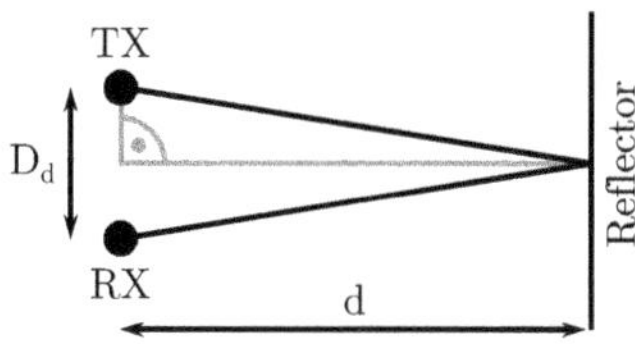

Figure 4.19: Calculation of the length difference due to the aperture and non parallel paths

For the calibration, the expected delays can be split into two different types: delays caused by different cable and fiber lengths and delays caused by the path the wave travels in free air. The paths traveled in free air to the reflector and back are not the same for all combinations of transmitter and receiver, since the reflecting plate is relatively close to the antennas, and the path forms a triangle with different sizes, depending on the position of the antennas involved.

Although the lengths of the cables and fibers in the synchronization are not precisely known, a difference in length of more than $\Delta l_{\text{sync}} = 10\,\text{mm}$ can be ruled out. Additionally, the microstrip feed lines of the patch antennas are of different lengths. The worst-case physical length differences are 14.2 mm and 27.7 mm on the transmit and receive side, respectively. This is a total electrical length difference of up to $\Delta l_{\text{el}} = 62.9\,\text{mm}$, when assuming a speed of $2/3c$ in the microstrip line.

The distance between the transmitter and the receiver induces an additional delay, compared to just twice the reflector distance, that needs to be taken into account. The problem is depicted in Figure 4.19. The reflector is usually placed about $d = 600\,\text{mm}$ away from the radar, while the worst case distance between any transmitter and receiver is $D_{\text{d}} = 219.9\,\text{mm}$. Therefore, the additional length difference is

$$\Delta l_{\text{path}} = 2\left(\sqrt{d^2 + \left(\frac{D_{\text{d}}}{2}\right)^2} - d\right) \approx 20\,\text{mm}. \tag{4.26}$$

The total worst case length difference is the sum of all three values

$$\Delta l_{\text{wc}} = \Delta l_{\text{sync}} + \Delta l_{\text{el}} + \Delta l_{\text{path}} \tag{4.27}$$
$$\Delta l_{\text{wc}} \approx 92.9\,\text{mm}. \tag{4.28}$$

This value is less than the maximum allowed length difference of $\Delta l_{\text{max}} = 195\,\text{mm}$, by a factor of approximately 2.1. This is sufficient for the measurements carried out, but one may consider placing the reflector at a larger distance and use length-matched feed and synchronization lines to reduce the delay differences, especially for even larger apertures.

4.3 System Parameters, Software and Signal Processing

The ideal radar for such an application would be a MIMO radar capable of transmitting simultaneously on all transmitters [64], especially if the system has a significant number of antennas. An FMCW based system, like the one described here, can only have one transmitter active at one time, as it is not possible to separate multiple transmitters from each other in the received signal. Nevertheless, the results of both system types are very similar. The main difference is that the FMCW-based system performs its measurements consecutively but assumes they are carried out simultaneously. This difference only manifests in the case of moving targets, as explained in Section 2.9. With no moving targets analyzed throughout this work, it is possible to treat both system types equally, and no movement compensation is necessary.

The parameters for the radar are chosen for a near- to mid-range scenario. The center frequency is $f_c = 78.768\,\text{GHz}$, corresponding to $\lambda \approx 3.8\,\text{mm}$ and the bandwidth is relatively high with $B = 1.535\,\text{GHz}$, translating to a range resolution of $\Delta R = 97.7\,\text{mm}$, according to (2.23). Each measurement has $N_{\text{ft}} = 256$ complex samples, sampled at $f_{\text{s,adc}} = 5\,\text{MHz}$, resulting in a total sweep time of $t_s = 51.2\,\mu\text{s}$ and a slope of $s = 29.9813\,{}^{\text{MHz}}/_{\mu\text{s}}$. The most important system parameters are summarized in Table 4.1 on page 55.

The data flow in the system is illustrated in Figure 4.20 and described in the following. The data from the radar arrives in four LVDS lanes, one for each ADC. These LVDS lanes operate at a rate of up to $600\,{}^{\text{Mbit}}/_{\text{s}}$ per lane, depending on the configuration of the RoCs. Additional lanes are responsible for the synchronization of the data, in which one lane is used to mark the start of a word (i.e., sample), and another marks the start of a frame (i.e., sweep). Once a set of samples from one of the two radars is received in the FPGA, they are packed into a 128-bit word. The data itself contains four samples of 16 bits each (64 bits), either I or Q data. I and Q data are transmitted alternatingly and are packed into the 128-bit structure independently.

The 128-bit word holds a start byte as the first byte, which is always 0xBB to identify the start of a word. This helps to resynchronize to the words once synchronization is lost. The second byte contains the channel identifier in the lower nibble and status bits in the higher nibble. The channel identifier is 0xA for data from the left radar, 0xB for the right radar, 0xC for command data, or 0xD for debugging data. The status bits are 0x4 for a start of frame and 0x8 for end of frame. These bits identify the first and last data word of a frame, which also helps to resynchronize in the case of lost or corrupted data. The following 8 bytes contain the four samples of data, followed by a word counter (4 bytes) and a frame counter (2 bytes), which help to keep track of the received data and notice if

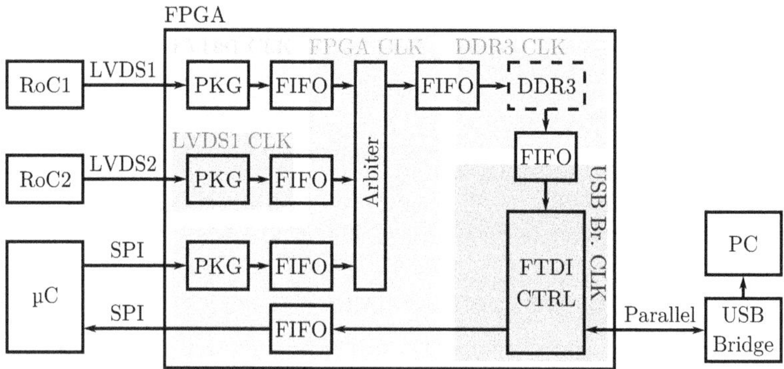

Figure 4.20: Data flow within the FPGA including multiple clock domains. PKG stands for packaging, CTRL for controller.

data is missing. The data counter is incremented for each word packed for this particular radar, and the frame counter increments once a frame is complete. All data is encoded as little-endian.

Once a complete word is received from an RoC and packed into a 128-bit word, it is buffered in a first in first out (FIFO) memory, one for each radar. These FIFOs are also used to cross clock domains within the FPGA from the LVDS clocks, which are provided from each radar on a separate LVDS lane, to the internal FPGA clock domain. An arbiter then combines the FIFO outputs from the LVDS streams into a single FIFO, which is made as large as possible to provide maximum buffer capability within the FPGA. Additionally, the external DDR3 memory can be used to increase the size of available buffer, which is only required for a large amount of data over an extensive time. The currently performed sweeps have enough idle time in between them, so the DDR3 memory is not required and therefore not utilized. Another FIFO is used to cross to the clock domain driven by the USB bridge. The data is transferred to the FTDI FT232H USB bridge using a 16-bit parallel bus, operating at 60 MHz.

The arbiter is supplied with an additional FIFO that holds messages from the microcontroller. These messages are coming from the SPI interface and are also packed into a 128-bit word, but only the first data byte is used. This simplifies the FPGA design, and the high data overhead is accepted, as the amount of data on these channels is negligible and even non-existent while doing measurements. The data coming from the USB bridge is also buffered in a FIFO and provided to the microcontroller.

On the PC, a piece of software, written in C++ for performance reasons, is used to connect to the radar modules using the USB interface. The software splits the words and frames into C++ structures and performs a sanity check using the various status bits and counters in the data stream. Once the dataset is verified, it is provided on a single transmission control protocol (TCP) socket. The TCP socket is a simple interface to the complete radar system, as most programming languages and other development environments have some libraries or built-in functionality to provide TCP connectivity. This also allows for easy remote control of the radar using standard networks.

The signal processing is implemented in Matlab. A custom library provides all the necessary functions to initialize the radar and load a particular sweep configuration. After the initialization, it is possible to trigger a measurement and receive the data from the radar with a single function call. Although the library and the rest of the system allow for live processing, the data is stored for post-processing. This off-line processing is also performed in Matlab.

First, a measurement for the calibration is carried out. As described in Section 4.2, a metal plate is used as reflector, which is placed roughly 600 mm away from and parallel to the radar's yz-plane. The measurement is saved and then processed using the algorithm described in Section 4.2. The resulting amplitude and phase errors are saved to a single file, for the measurement processing script.

Next, a measurement of a target scenario is carried out. A phase-shift-and-sum beamformer generates images of this scenario in front of the radar. According to (2.25) and the largest virtual aperture in a single axis of $D_{\mathrm{r}} = 133\,\mathrm{mm}$, the far-field distance is approximately 9.4 m. Therefore, the near-field beamforming matrix from (2.31) is appropriate, since the radar will operate in near-field conditions. Additionally, (2.35) performs a scan in both azimuth and elevation.

After applying the beamforming, the time axis is weighted with a Hamming window to reduce side lobes. [10, p. 33] implies that this will deteriorate the range resolution by a factor of 1.3. The expected range resolution of the system including the window function is therefore $\Delta R_{\mathrm{h}} = 127\,\mathrm{mm}$ instead of 97.7 mm. An FFT then transforms the time axis into the range axis.

The result is a three-dimensional image of the scenario, given in azimuth, elevation, and range. Although a scan of azimuth and elevation is performed for a particular focus distance, the results can also be plotted along the range axis, resulting in slightly reduced sharpness for targets further away from the focus distance.

In addition to the evaluation of measured data, there are scripts to generate simulated data. These scripts rely on the same parameters as the other scripts, for example, the position of the antennas or the sweep parameters. They generate data in the same output format as the measurement script and can easily replace a measurement with the according simulation. There are two different simulation scripts. The first one performs a simulation of the reflector wall to generate calibration data. There is theoretically no calibration data required because there should be no errors in a nearly perfect simulation. However, the simulation of the calibration scenario helps verify the calibration script, as the result is already known. Additionally, it simplifies the beamforming script, since no differentiation between simulated and measured data is necessary.

The second simulation script generates the data for a point target. This script is used to compare its results to the measurement of a corner reflector target. Both simulations are based on the Matlab Phased Array Toolbox and use the same system parameters given in Table 4.1. The toolbox is mainly used for generating the transmitted signal and delaying it. However, the simulation also includes features such as the antenna gains, the receiver noise figure, an RCS-based model of the target, and proper power levels.

4.4 Simulations, Measurements and Results

To validate the system, the measurements are compared to simulations of the same scenario. The system is placed in an anechoic chamber, as depicted in Figure 4.14 on page 66. The target is a small corner reflector of length $L = 18\,\mathrm{mm}$. It is placed at a distance of approximately 2.3 m in front of the radar. The stand holding the reflector is covered with anechoic material and has a roughly 50 cm horizontal boom towards the radar, holding the reflector at its end. This reduces the reflections from the stands. The reflector is easily distinguished from the stand owing to their difference in range. Both simulation and measurement are processed with the same Matlab script from the previous section. The script performs a scan in azimuth and elevation.

Figure 4.21 shows the top view of the scenario with a cut through the maximum target amplitude. The amplitude is normalized to the maximum value. The target can clearly be identified slightly right of the boresight axis at a distance of 2.3 m, as expected. To the left and right of the target, the side lobes create the expected clutter, caused by the radiation pattern. The very weak reflection at a distance of about 3 m is likely from the stand, which is detected despite being covered with anechoic material. The clutter closer in than the target, at around 1.2 m, is probably caused by reflections from the floor.

Figure 4.22 depicts the same measurement as the one in Figure 4.21 but in the azimuth/elevation plane at the target's distance. The amplitude is also normalized to the maximum value. In this view, the target can also be identified in the center, with side lobes surrounding it. In comparison, Figure 4.23 shows the simulated result of the same scenario. The results look very similar, although certain differences are visible in the details. For a better side-by-side comparison, a cut in the azimuth plane of both is made and plotted in Figure 4.24. The main lobe and the first side lobe on each side are almost identical. The overall shapes are comparable, with the local maxima roughly at the same positions, but the amplitudes of the peaks differ. The overall side lobe level is also similar to the simulation. There are several possible explanations for the differences seen:

- Certain effects of the radiation pattern of the individual patch antennas are not corrected. Of course, the effect of the radiation pattern is included in both calibration and measurement. However, the calibration is executed relatively close to the radar compared to the measurement. Therefore, the beam angles differ, and both phase and amplitude may vary between the two angles. Additionally to the radiation pattern of the patch itself, local effects, like coupling between antennas, or reflections on close objects, like the radar chips, will influence each patch individually. To compensate for this effect, the calibration needs to be more sophisticated and including a phase and amplitude radiation pattern of each patch antenna. These would have to be measured on the radar itself, introducing a new topic to consider. As such a complex calibration is not the objective of this demonstrator, the negative effects are accepted.
- Several potentially negative influences can be summarized under the term mechanical tolerances. There is, for example, the positioning of the modules relative to each other. With such a short wavelength of 3.8 mm, precise positioning is crucial. Care has been taken, but a misalignment of several tenths of millimeters is possible. There are also tolerances in the manufacturing process of each patch antenna, which may influence the phase center (i.e., the position) or the radiation pattern (i.e., amplitude and phase) of the patch. Those tolerances will affect the overall radiation pattern of the array. Finally, the calibration plate is an aluminum metal sheet of approximately 600 by 600 mm. It is for sure not perfectly flat, which may influence the phase of the calibration, depending on where the reflection for each transmitter/receiver combination occurs.
- The corner reflector has a size of $L \approx 18\,\mathrm{mm}$. The minimum distance based on the HPBW of the reflector is calculated from this value using (4.5) and yields approximately 2.9 m, which is very close to the actual distance of 2.3 m. Although the condition still holds for these values, it means that there is almost a 3 dB drop in amplitude for the two antennas that are the furthest away from each other. This effect is not accounted for and may show up as amplitude errors.

The measured angular resolutions have the value of $\Delta\gamma_{\mathrm{meas}} = 1.37\,°$ and $\Delta\delta_{\mathrm{meas}} = 1.67\,°$ for azimuth and elevation, respectively. The simulated values from Table 4.1 of 1.40 ° and 1.64 ° match almost perfectly. The measured range resolution including a Hamming window $\Delta R_{\mathrm{h,meas}}$ is 130 mm, which also perfectly fits the calculated value of $\Delta R_{\mathrm{h}} = 127\,\mathrm{mm}$.

Using a different arrangement of the modules, the system has also shown an angular resolution as low as 0.6 °, presented in [60]. In this case, the antennas of three modules were placed along the horizontal axis, providing only azimuthal resolution. Each module provided eight receivers and three transmitters to the measurement, resulting in a total virtual aperture of 450 mm. As this system does not provide any resolution in elevation and does not contribute any additional insight besides the better angular resolution, it is not shown here.

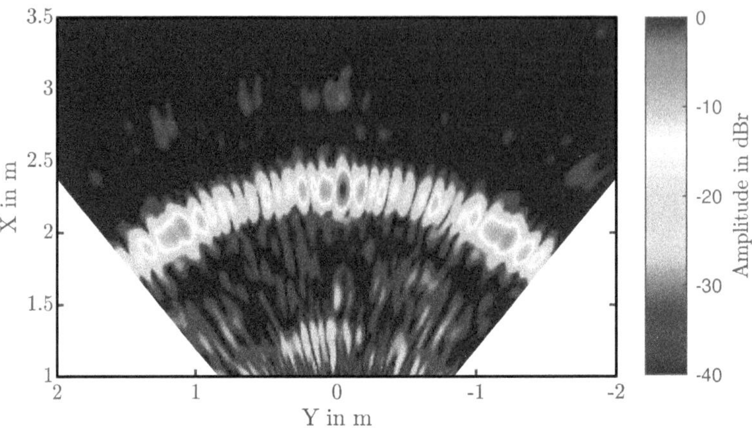

Figure 4.21: Measurement result in top view of a corner reflector at 2.3 m using the synchronization demonstrator

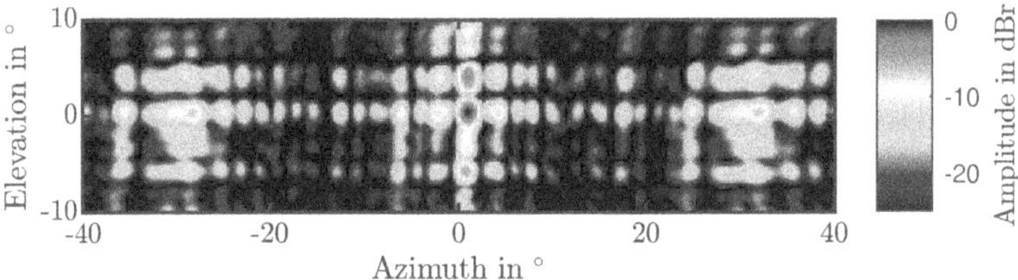

Figure 4.22: Measurement result in frontal view of a corner reflector at 2.3 m using the synchronization demonstrator

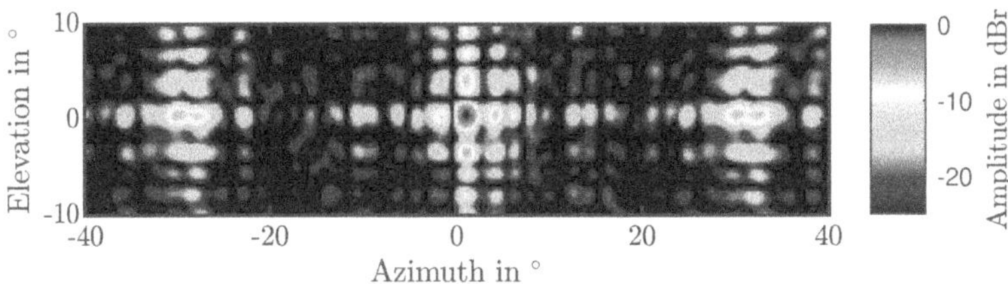

Figure 4.23: Simulation result in frontal view of a corner reflector at 2.3 m using the synchronization demonstrator

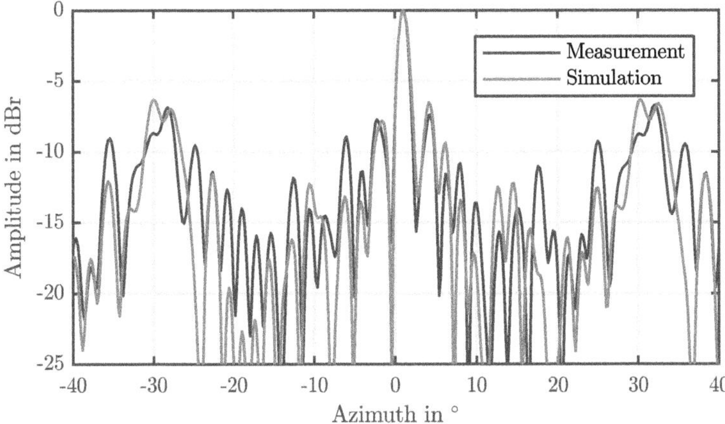

Figure 4.24: Comparison of simulation and measurement of a corner reflector target at 2.3 m using the synchronization demonstrator

4.5 Summary

The demonstrator had three primary goals: to show a modular radar system, to achieve synchronization using a radio over fiber system, and to exhibit an adequate angular resolution compared to current state of the art automotive radar systems. All three goals were fulfilled and the system performed as expected. The modules were arranged in different configurations and demonstrated the modularity of the system. The synchronization worked flawlessly, allowing a coherent operation of all modules to form a modular radar with a large aperture. The measured and simulated scenario of a single corner reflector showcased the capabilities of the system and the proposed array arrangement for a radar system with high-resolution beamforming in azimuth and elevation. Simulation and measurement of the corner reflector matched well. Although the angular side lobe level is relatively high to the advantage of the good angular resolution, a single target could easily be detected. The angular resolution reached as low as 1.37 ° and 1.67 ° for azimuth and elevation, respectively, which far exceeded that of commonly available automotive systems. The system also demonstrated a resolution down to 0.6 ° in another work.

5 SAR Demonstrator

The first demonstrator showed the concept of a modular radar system, which can operate coherently through an RoF synchronization. It also demonstrated a high angular resolution radar system, but the antenna array was primarily optimized for the angular resolution, leading to an inferior SLL. The L-shape of the array additionally limited the number of usable channels as each module used either the transmitters or the receivers, but not both of them.

The second demonstrator, introduced in this chapter, shows the antenna arrangement proposed in Chapter 3. This array features a far better SLL and an even better angular resolution. The demonstrator is used for a mid- to far-range configuration in a much larger and more realistic scenario for an automotive radar. Besides the improved antenna array, this demonstrator is also used to measure the same scenario as the first demonstrator in the anechoic chamber, and a comparison between the two measurements is made.

5.1 Concept and Hardware

To keep this demonstrator's cost and complexity as low as possible, it uses a synthetic aperture instead of building a system with 48 transmitters and 64 receivers. The concept of a synthetic aperture was introduced in Section 2.13. In this case, synthetic aperture means the usage of a single transmitter and a single receiver only. Measurements are taken at different positions and combined in software afterward to form the synthetic aperture. The following sections will describe the utilized hardware and its capabilities.

5.1.1 Synthetic Virtual Aperture

Of course, it is possible to form a synthetic aperture using the transmit and receive positions of the array to model. The transmitter would be positioned at the first location; then, the receiver would be placed at all receiver positions of the array, one after another. Next, the transmitter position is changed, and the next set of receiver positions is measured until all combinations are done. The problem here is the very complicated arrangement of the transmitter and receiver. Both have to be able to be positioned arbitrarily without physically interfering with each other. This is almost impossible, as transmit and receive elements can be placed as close as $\lambda/2$ next to each other in any orientation.

To overcome this issue, the demonstrator uses not just a synthetic aperture but also the virtual aperture formed from all transmitters and receivers. The virtual aperture, introduced in Section 2.14, is not just a trick to simplify the signal processing for the array. A virtual array physically sampled at its positions has the same properties as the physical array used to construct the virtual array, at least under far-field conditions. As described in Section 2.14, it is possible to position a single transmitter and a single receiver next to each other and sample the mono-static virtual aperture with it. The mono-static virtual aperture is the standard virtual aperture scaled by 1/2. This is required because both the transmitter and receiver are moving.

The drawback of this procedure is that the measurement of each transmitter/receiver position combination is carried out successively. Between each measurement, the antennas have to be repositioned, which takes some time even when automatized. Therefore, it is not possible to measure moving targets with this approach easily as the target is assumed to not move between the measurements. Easy compensation of the movement would only be possible if the target moves less than one range-bin, as explained in Section 2.9. This is not to be expected, as the SAR measurements carried out within this chapter require about one hour for a complete scan.

5.1.2 Radar Module

The radar used for this demonstrator is an off-the-shelf evaluation board for the Texas Instruments AWR1642 RoC [65] called AWR1642BOOST [66], depicted in Figure 5.1. The AWR1642 is very similar to the AWR1243P, which was used for the first demonstrator. However, there are a few differences between the two chips. The AWR1642 has only two transmitters, compared to the three transmitters of the AWR1243P. It does not support synchronization between multiple chips, which is not required in this case. The data can be stored internally, and the integrated ARM Coretex-R4F microcontroller can be used for signal processing. The typical use case for this RoC is to completely process the measurements within the chip, including target detection and AoA estimation. The only output is a list of targets, usually provided over the integrated controller area network (CAN) interface in case of an automotive application.

However, the software running on the internal microcontroller is custom made for a particular application and can be utilized to access the raw measurement data. The raw data is required for this application because the beamforming has to be carried out after all measurements are done. Additionally, the internal memory would not be enough to store all measurements, and off-line processing of the data is preferred. The microcontroller in the RoC has several interfaces that can be utilized to stream the data to an external PC. One of the universal asynchronous receiver transmitter (UART) connections is routed through the USB connection and is thus the easiest way to access the data. A UART connection is not particularly fast, but the radar has to be moved mechanically between each measurement. The time during the movement and settling phase can be used to transfer the data to the PC. The microcontroller software is based on an example provided by Texas Instruments along with the software framework. It already allows the configuration of the radar with a configuration provided over the UART link. A few modifications enable the software to transmit the raw data instead of a target list back to the PC.

Figure 5.1 depicts a photograph of the evaluation board. On the right-hand side are the four receive antennas in a group at the top and the two transmit antennas with more distance to each other at the bottom. The receive antennas are guarded with an additional antenna on each side. The module is mounted with a rotation of 90° counter-clock-wise, providing vertically polarized antennas. The antennas are serially-fed arrays with four elements each. According to [66], each array has an HPBW of approximately 70° and 30° for azimuth and elevation, respectively, once rotated. The reduced HPBW in elevation, compared to an individual patch, is not critical for the following applications. When looking at real scenarios on the road, a limitation in elevation is usually not critical, as most targets of interest are located close to zero-elevation. A serially-fed array with a few elements might

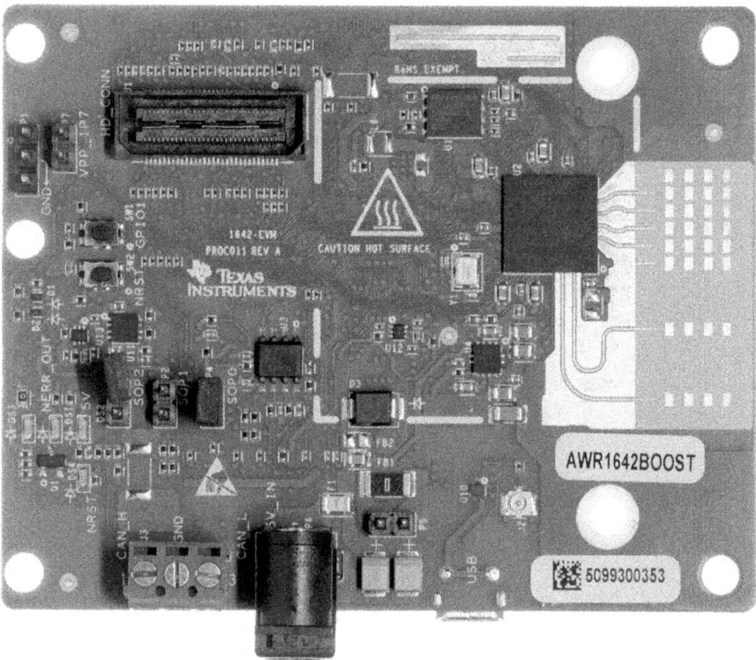

Figure 5.1: Photograph of the Texas Instruments AWR1642BOOST evaluation board

even be beneficial to increase the antenna gain and, in consequence, the maximum range. According to [66], the antennas have a gain of at least 9 dBi, 4.5 dB higher than the single patch used for the first demonstrator. Applied to transmitter and receiver an additional gain of 9 dB would be possible, leading to an increase in range by 68 %.[1] Of course, the optimization of the antenna positions in Chapter 3 would require some minor modifications to account for the larger antenna elements. As the array is sampled as a synthetic aperture, the antennas' overlap is not a problem and has no major influence on the result, except for the higher gain and slightly limited elevation angles.

[1] 9 dB is a power factor of 7.9. The range in (2.19) is proportional to the fourth root of the power-gain, resulting in $\sqrt[4]{7.9} = 1.68$, corresponding to an increase by 68 %.

5.1.3 XY-Table

The radar module is mounted on an XY-table, allowing to position the radar within a two-dimensional plane of 600 by 600 mm with a theoretical resolution of 2.5 µm. Each axis runs on two stainless steel 25 mm linear guides, which are mounted on an aluminum frame of 45 mm profiles. The massive components are required to ensure a stiff construction, as relative movements should stay far below the wavelength of approximately 3.8 mm. The vertical axis is driven by two stepper motors, while the horizontal axis uses only a single stepper motor. All three motors are driven and controlled from a central unit. The microcontroller is connected to a PC using USB and expects standard G-code commands to control the two axes. A switch at the end of each axis is used to align the position after a reset or loss of power. Figure 5.2 shows a photograph of the XY-table, including the mounted radar module.

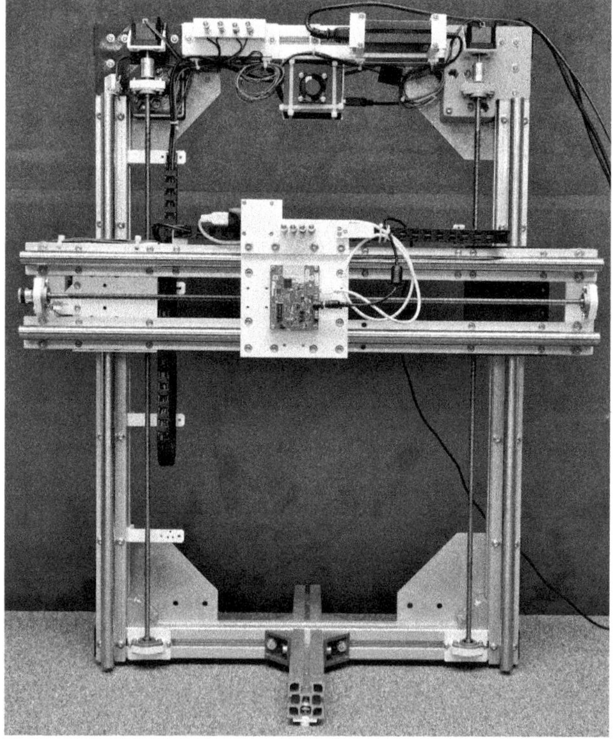

Figure 5.2: XY-table with the mounted radar module

5.2 Software and Signal Processing

The radar and the XY-table are controlled using a single Python script. The table is moved to a position, and the script ensures that the position is reached by polling the status flags of the GRBL[2] firmware, which is controlling the XY-table. Once the position is reached, a measurement is triggered. The radar firmware reports the completion of the measurement, which is used as an event to move to the next position. During that movement, the measured data is transferred to the PC. The process then starts again for the next position. The whole system is optimized for low latency and fast movement of the XY-table. When scanning a complete grid, all horizontal positions for a given vertical position are measured in a row, as the horizontal axis can move faster due to its lower mass. With these optimizations, it is possible to measure up to four positions per second.

The measured data is stored in a single file for off-line processing. The file format is the same as the one used for the first demonstrator. This allows using the same signal processing scripts like the one used for the first demonstrator, described in Section 4.3. It also makes the comparison of two data set easy as both can be processed with the same script.

The signal processing is extended by a target detection using the CFAR algorithm. It is used to clean the three-dimensional result space and detect only true targets. This makes it possible to plot the result in a three-dimensional figure, and it also clarifies what such a radar system would actually detect. The resulting target list may then be used for target classification. It includes the position in three dimensions, the reflectivity, and the Doppler information in case of a CS radar.

5.3 Measurements

This demonstrator is used for two different measurements. Firstly, it is used to measure the same scenario as the first demonstrator did, and the results are compared. Secondly, it measures the proposed antenna array from Section 3 to demonstrate the capabilities of such a radar system.

5.3.1 Synchronization Demonstrator Scenario

The first measurement is carried out under very similar conditions compared to the measurement described in Section 4.4 with the synchronization demonstrator: A single corner reflector target with a size of $L = 18\,\mathrm{mm}$ is placed in a distance of 4.81 m within an anechoic chamber. The antenna configuration is the same, except it is using the mono-static virtual aperture instead of the physical aperture from Section 4.1.4. The virtual array is constructed by the convolution of the transmitter and receiver array, as described in Section 2.14. The sweep parameters are also identical to the ones before. A bandwidth of $B = 1.535\,\mathrm{GHz}$ at $f_c = 78.768\,\mathrm{GHz}$ with a slope of $s = 29.9813\,\mathrm{MHz/\mu s}$ is used. The ADC is capturing $N_{ft} = 256$ samples with a sampling rate of $f_{s,adc} = 5\,\mathrm{MHz}$. Consequently, the other system parameters, like the resolutions in azimuth, elevation, and range, are the same.

Figure 5.3 shows the result of the measurement compared to a simulation of the same scenario. The figure can be compared to Figure 4.24 on page 79. It shows a cut in azimuth with fixed elevation and a focus distance right on the target's reflection. Compared to

[2] Documentation and software is available here: https://github.com/gnea/grbl/

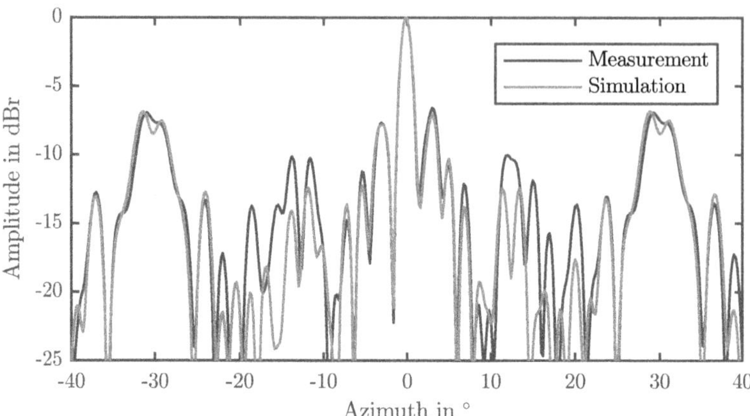

Figure 5.3: Comparison between the measurement and the simulation of the SAR demonstrator with the antenna configuration from Chapter 4

Figure 4.24 the result in Figure 5.3 fits its simulation better. There are two reasons for that: The target is slightly further away from the radar, reducing the problem with the too-narrow beam from the corner reflector, discussed in Section 4.4. The second reason is the non-existent mutual coupling between adjacent array elements. As the aperture is sampled virtually, there is only one transmitter and one receiver present at any given time and it is the same for all measured positions. Additionally, the SAR based system does not have different antenna elements, and no calibration is required, reducing the system's potential errors even further.

5.3.2 Outdoor Scenario

The second measurement covers a more realistic outdoor scenario. The scenario is shown in Figure 5.4. The radar is located at the bottom left corner of the picture pointing parallel to the water drain. Figure 5.5(b) shows the scenario in the top view. The radar is placed 5 m below the bottom of the shown overview, i.e., at the position 0/0 in Figure 5.5(a). The scenario consists of a wall with windows to the left (B), a metal and wood box (D), a metal fence (E), a bike (C), a car (A), and several other objects.

For this scenario, the radar is configured for a mid- to far-range application. As there are no moving targets, an FMCW configuration is used. Moreover, the radar would not be able to perform the measurements required for a CS modulation because the measurements are not performed simultaneously for all transmitters and receivers. The bandwidth is 768 MHz, resulting in a range resolution of 20 cm according to (2.23) or 25 cm when applying a Hamming window. The scenario has a maximum range of approximately 22 m. Using (2.13) results in a minimum number of samples of

$$N_{\mathrm{st,min}} = \frac{4 d_{\mathrm{min}} B}{c_0} = 220.$$

Usually it is advantageous to use a power of two for faster signal processing. Therefore, a value of $N_{\mathrm{st}} = 256$ is chosen. The ADC is sampling at 5 MHz, resulting in a ramp duration of 51.2 µs, just like in the earlier scenarios. The slope is

$$s = \frac{B}{t_{\mathrm{s}}} = \frac{768\,\mathrm{MHz}}{51.2\,\mathrm{\mu s}} = 15\,\mathrm{MHz}/\mathrm{\mu s}.$$

Like before, a scan in azimuth and elevation is performed using the near-field beamforming approach from Section 2.8. The focus distance is set to 15 m, being in the center of the scenario. The results of the beamforming are shown in Figure 5.5(a) for an elevation of 0 °. The amplitude is scaled to the maximum target amplitude and plotted logarithmically. The radar's $z = 0$ position is approximately 52 cm above the ground.

The beamforming result in Figure 5.5(a) clearly shows all relevant targets. The reflections from the fence (E) are well visible; even the steps in the fence are readily identifiable. The back of the car (A) is visible, but not the sides, which is expected. Even the bicycle (C) is detected. The wall (B) has a few strong reflections, especially at the frequently spaced vertical metal separators. The box (D) and also the air conditioning system behind it are identifiable. However, there are two reflections that may be considered false positives. The first one is (F), which can be explained easily: It is the fence, which is reflected from the car's side. The reflection on walls, targets, and other objects is a known problem in radar systems. The second unknown signal is (G), which cannot be explained by a reflection as there is no potential reflector towards the indicated target. The only possible explanation is a reflection from the meshed water drain, which is much more visible for lower elevation values. The round circles which are spread all over the image are caused by the side-lobes of strong targets. They are always circles around the center of the radar, as side-lobes spread across azimuth and elevation, or in other words, they have a constant range.

Figure 5.6 shows two more beamforming results for the same scenario but different elevation angles. The left image has an elevation of 3 °, the right image has 6 °. Most of the targets are not visible anymore, as the beam points above them. The car and the fence are only visible for 3 ° elevation. On the other hand, the wall has much more reflections in both images, compared to 0 ° elevation. These are the windows, which were not visible for zero elevation.

To detect individual targets in the beamforming result, an CFAR algorithm is utilized. The algorithm itself is described in Section 2.15. It is clear from the previous analysis of the antenna array, but also from the beamforming results in Figures 5.5 and 5.6, that the result has fairly high side lobes along azimuth and elevation compared to the range axis. Therefore, the CFAR algorithm only uses a two-dimensional cut in azimuth and elevation for its training and guard cells. This way, it will use the higher side lobes as a reference, which is the true noise plus interference level the cell under test should be compared to. The beamforming is carried out with a resolution of 0.5 ° in both azimuth and elevation. The CFAR algorithm uses six guard cells and 14 training cells in each axis. The three guard cells correspond to an angle of 1.5 °, which is a bit higher than the expected angular resolution. This ensures that the signal does not interfere with itself. The number of training cells is chosen a couple of times higher to get a good average of the surrounding noise and interference level. The minimum SNR for a target to be detected is chosen to be 13 dB.

The result of the CFAR algorithm is shown in Figure 5.7 in a three-dimensional plot. A circle indicates each detected target. The color of the circle additionally indicates the height above zero-elevation. The xy-plane is referenced to the radar system, thus a height of 0 m is

Figure 5.4: Scene for the SAR measurements with the radar at the bottom left of the picture

in fact 52 cm above the ground. The plot only shows targets for an elevation of at least 0 °. This cleans the plot significantly as reflections from the ground are not shown. In a practical application, the ground would have to be detected by the following signal processing and categorized as such. All major targets are again marked with the corresponding letter.

The wall (B) is clearly visible, and it is even possible to identify individual window frames. The car (A) still has the primary reflections at its back, but there are two additional reflections further to the front of the car. The lower (blue) one probably comes from the front tire or the wheel case, while the mirror or the A-pillar may cause the higher (cyan) reflection. The bicycle (C) is even clearer visible than before.

5.4 Summary

This chapter introduced the usage of an XY-table to build a synthetic virtual aperture radar system. This radar is very versatile, as it can be utilized for different scenarios and antenna configurations. An off-the-shelf radar evaluation board is used for this demonstrator, which is programmed to transfer the raw radar data over the integrated USB interface to a PC for off-line processing. The data is processed using the same scripts as the first demonstrator to easily exchange the data between the two systems.

The first measurement used the SAR demonstrator in combination with the antenna array arrangement of the first demonstrator. The results showed excellent agreement with the simulation, better than the measurement of the first demonstrator. This proved that parasitics and, in general, the difference between individual antennas are mostly responsible for the discrepancy between simulation and measurement visible in the measurement with the first demonstrator, shown in Figure 4.24 on page 79.

The second measurement used the array proposed in Chapter 3, which provides a much better SLL and even better angular resolution. With the utilization of a CFAR algorithm, it was possible to reconstruct a good three-dimensional image of a complex outdoor scenario

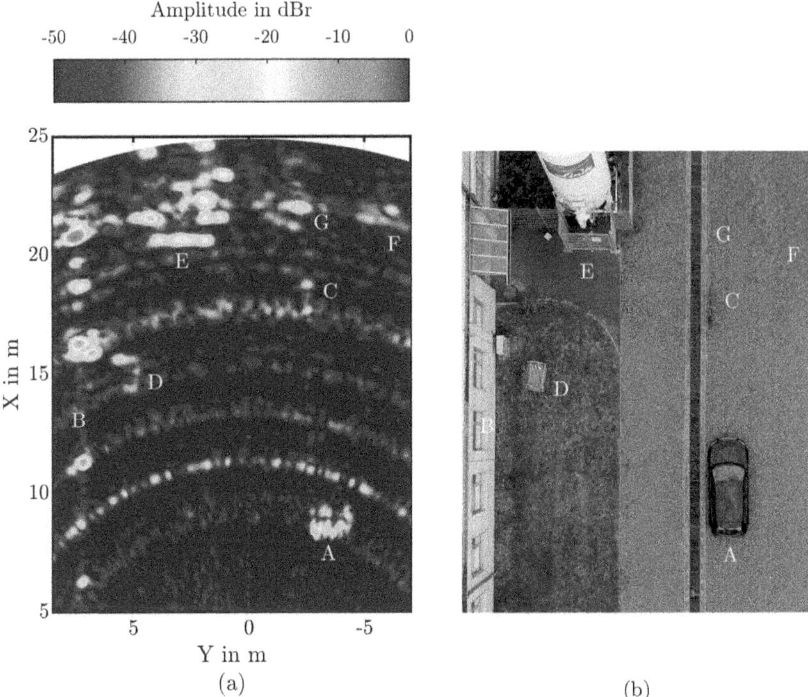

Figure 5.5: Beamforming results (a) and a photograph (b) in top view of the SAR demonstrator for the outdoor scenario with 0 ° elevation and the following targets: car (A), wall (B), bicycle (C), metal/wood box (D), fence (E), fence ghost target reflected of car (F), and meshed water drain (G)

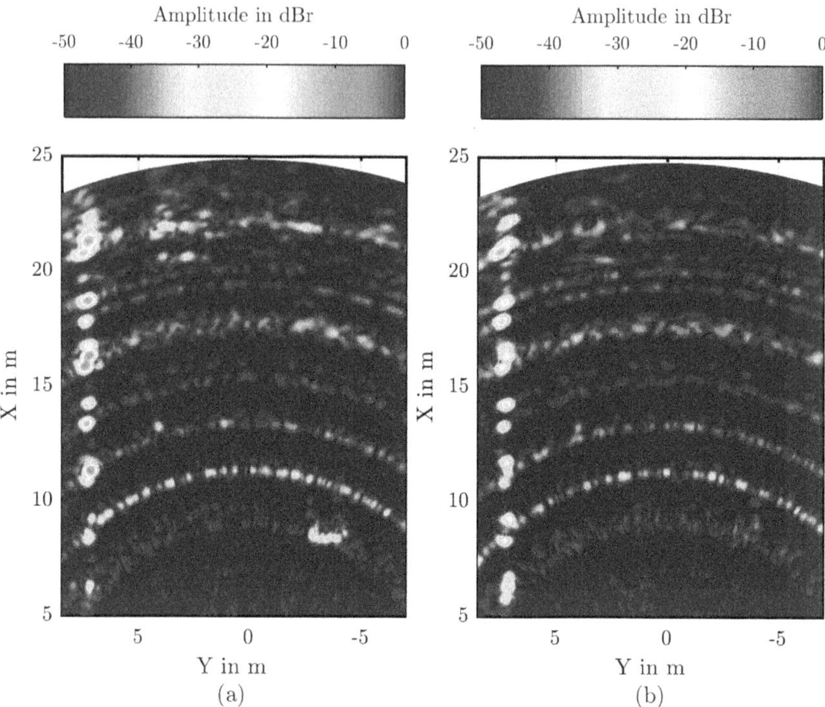

Figure 5.6: Beamforming results in top view of the SAR demonstrator for the outdoor scenario with 3° (a) and 6° (b) elevation

featuring different types of targets. Overall, the SAR radar demonstrator performed as expected, even in a realistic environment.

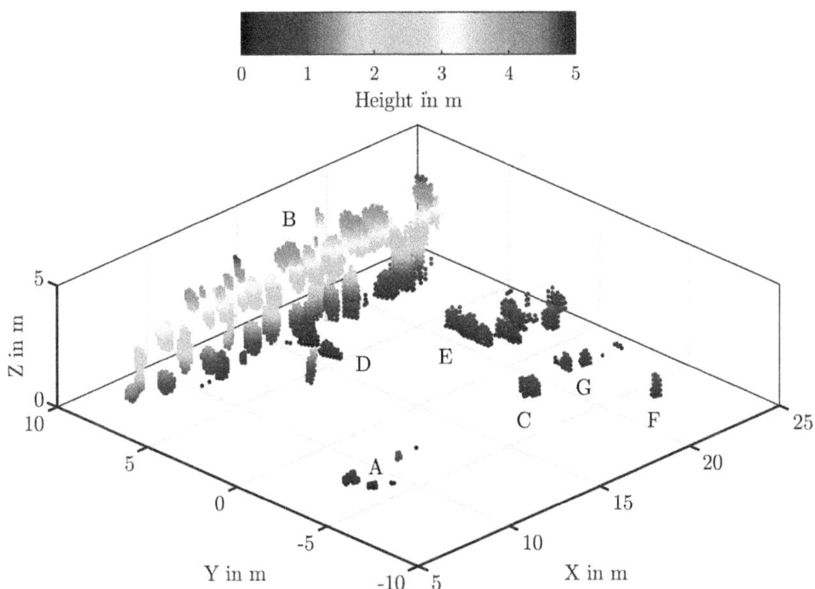

Figure 5.7: Three-dimensional image of the target detection using CFAR. The color additionally indicates the hight of the target relative to the radar from 0 m (blue) to 5 m (red).

6 Module Position Tolerance Compensation

Many effects and external influences have not been analyzed or did not affect the two demonstrators built in the last two chapters. However, one influence that was already mentioned did affect the presented systems and will also be a problem in the case of a mass-produced radar system: the antennas or modules' actual position relative to each other compared to the design positions. This chapter will look at typical causes for the displacement of the antennas, estimate the potential displacement, and the effect on the radar system. Finally, a software-based on-the-fly calibration system to compensate most of the system degradation is proposed and evaluated with simulations.

A similar calibration approach is presented in [67]. The drawback of this approach is that it requires very specific measurements in a laboratory environment and cannot run during the system's operation. Another similar approach is described in [68], but in this case, the calibration is not used to correct the antenna positions. Instead, offsets within the system, like length differences in cables and feed lines, are compensated. These differences are assumed to be known for the following calibration method as they do not depend on manufacturing tolerances as heavily as the antennas' position and can be compensated with a single measurement of a known target, either during production or during operation based on the help of other sensors.

6.1 Causes and Displacement Estimates

There are a few influences that may cause the antennas to be located slightly incorrectly. Especially the position of the modules relative to each other is susceptible to tolerances. There are two generally different causes for the modules' displacements: The first one is caused by manufacturing and mounting tolerances, which will stay constant after production. The second type of tolerances are those happening during operation like thermal expansion.

Theoretically, the manufacturing and mounting tolerances can be reduced to an arbitrarily small value but causing much higher manufacturing costs, lower yield, or both. To prevent displacement issues that emerge during operation, the options during manufacturing are limited, at least. In the case of thermal expansion, the right materials may help to reduce the problem, but it is not possible to completely eliminate the effect.

For the following analysis of the effect of module displacement, a few assumptions on the expected displacement are made. The antenna array proposed in Chapter 3 has a physical horizontal aperture of 144.3 mm. This is used as the base for worst-case mechanical tolerances. According to ISO 2768-1, even the fine class allows a tolerance of up to 200 µm for the given nominal size range [69]. If the module is a PCB mounted on a metal frame, this tolerance applies to both the frame and the board, summing to a worst-case value of 400 µm. This value is used as a reference value for the maximum constant module displacement, although it does not even include all mechanical tolerances. Others may be the tolerance of mounting the board to the frame: If the diameter of the screw is smaller than the diameter of the hole in the PCB, the position of the PCB may be shifted by up to half the diameter difference.

The tolerances changing during operation are estimated with the thermal expansion of the antenna. The automotive industry has a common temperature range of -40 °C to 125 °C for electronic components. If the radar is designed for the center temperature of 42.5 °C between these two extremes, a maximum deviation of 82.5 K is possible. Table 6.1 lists common materials, their thermal expansion coefficient, and the expansion of a 144.3 mm long element by a temperature change of 82.5 K. When assuming an aluminum frame for the radar, an additional expansion of 275 µm has to be expected.

Table 6.1: Thermal expansion examples for typical materials and the proposed array from Chapter 3

Material	Thermal expansion coefficient	Expansion for 144.3 mm and 82.5 K
Rogers RO3003 [61]	$17 \cdot 10^{-6}\,\mathrm{K}^{-1}$	202 µm
Stainless steel [70, p. 12-224]	$17.3 \cdot 10^{-6}\,\mathrm{K}^{-1}$	206 µm
Aluminum [70, p. 12-206]	$23.1 \cdot 10^{-6}\,\mathrm{K}^{-1}$	275 µm
Polypropylene [71, p. 252]	$150 \cdot 10^{-6}\,\mathrm{K}^{-1}$	1786 µm

6.2 Impact on the System

When assuming an Aluminum frame on which the modules are mounted with a tolerance of 400 µm in the y and z axes and the maximum displacement due to temperature of up to 275 µm, a maximum displacement of 675 µm has to be expected. The effect of a random module displacement of up to ±675 µm in the two axes is shown in Figure 6.1. The figure is based on 10000 simulations with the array proposed in Chapter 3, each calculating the received signal for a single target at an azimuth angle of 40 ° with the actual (displaced) module positions according to the channel model in (2.26). Then, beamforming is applied with the ideal (non-displaced) antenna positions. All beamforming results are normalized to the target amplitude if no displacement is present. The results are then combined by calculating the probability density function for each angle value. The color visualizes the probability density. The resolution in azimuth is 0.1 °, and the resolution for the amplitude is 0.2 dB, which is used for the probability density calculation. The simulation does not include any noise in order to analyze the impact of the antenna displacement itself.

It is obvious from Figure 6.1 that the unknown displacement of the modules has a negative influence on the side lobe level. Depending on the actual displacement of the antennas, the SLL can increase by a few dB. Especially the side lobes close to the target's main lobe are affected by unknown tolerances. The main lobe at $\varphi = 40\,°$ stays almost identical, but a loss in the directivity of one or two dB has to be expected.

Besides the increased SLL, the target's peak angle deviates from the optimal value. Figure 6.2 shows the distribution of peak azimuth angles for the previous example. In this case, the standard deviation is approximately 0.16 ° and the worst-case offset is below 0.5 °. This offset may not be considered as critical as the increased side lobes.

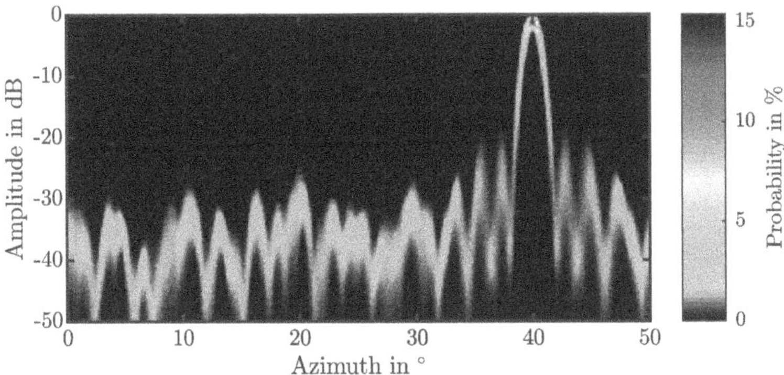

Figure 6.1: Effect of unknown random module displacement of up to ±675 µm for a single target from 40 ° azimuth, and zero elevation at 79 GHz

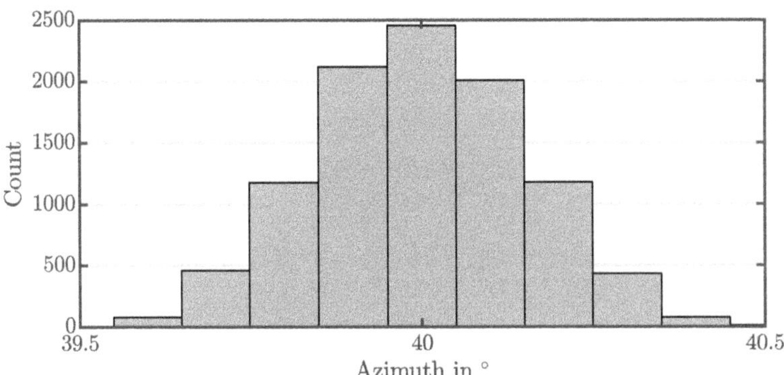

Figure 6.2: Effect of unknown random module displacement of up to ±675 µm for a single target from 40 ° azimuth, and zero elevation at 79 GHz on the peak angle

The effect of increased side lobes becomes even more problematic for systems operating at a higher frequency as the absolute displacement due to the mounting tolerances stays the same and does not scale with the antenna's size. The same scenario as before but at 240 GHz is shown in Figure 6.3. The maximum displacement due to the mounting tolerances is still 400 µm, while the thermal expansion displacement is scaled to 91 µm, as the aperture and thus the thermal expansion is proportional to the wavelength. Therefore, the simulation has a total maximum random displacement of 491 µm.

The main lobe is now significantly impacted, both in its amplitude and its azimuth angle. The distribution of the peak angle is shown in Figure 6.4. The standard deviation is now approximately 0.58 ° and the worst-case offset is 1 °, which might influence the system's performance significantly. The side lobes are even higher; those close to the main lobe can reach values above -10 dB relative to the optimal main lobe.

As only the y and z axes' displacement is analyzed, a target on the boresight axis is not negatively affected if far-field conditions apply. In this case, the received signal would arrive at all receivers simultaneously (i.e., with the same phase), independent from the module positions. The effect gets continuously worse for targets further away from the boresight axis. This means that the side lobes tend to increase even further in the given examples from Figure 6.1 and 6.3 if the target is located further away from the boresight axis.

The question arises whether the knowledge of the actual displacement would improve the side lobe level and the impact on the main lobe. Figures 6.5 and 6.6 contain the same simulations as the one from Figures 6.1 and 6.3 at 79 GHz and 240 GHz, respectively. The difference is that these simulations use the actual antenna positions for the beamforming step instead of the ideal positions. The resulting side lobe levels are much less spread. However, they are not completely identical for all displacements. This is expected, as the array changes slightly due to the module displacement and thus the radiation pattern is changed. Nevertheless, the resulting radiation patterns are much better compared to the ones with unknown displacement. The main lobe is completely restored in both cases and shows no spread in the target's angle.

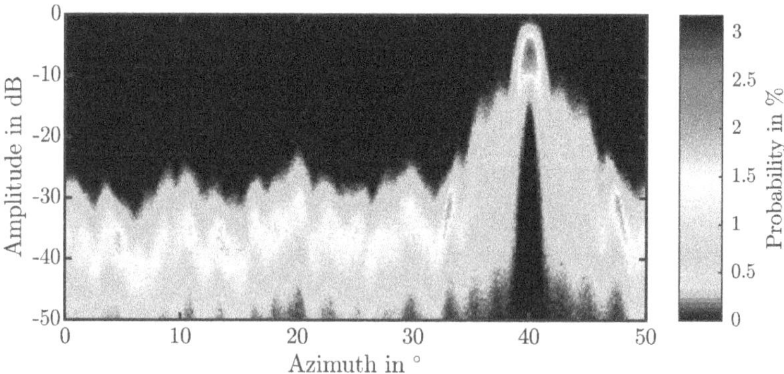

Figure 6.3: Effect of unknown random module displacement of up to ±491 µm for a single target from 40 ° azimuth, and zero elevation at 240 GHz

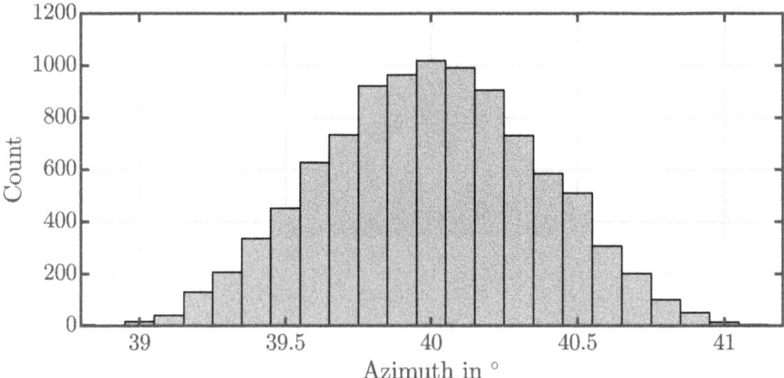

Figure 6.4: Effect of unknown random module displacement of up to ±491 µm for a single target from 40 ° azimuth, and zero elevation at 240 GHz on the peak angle

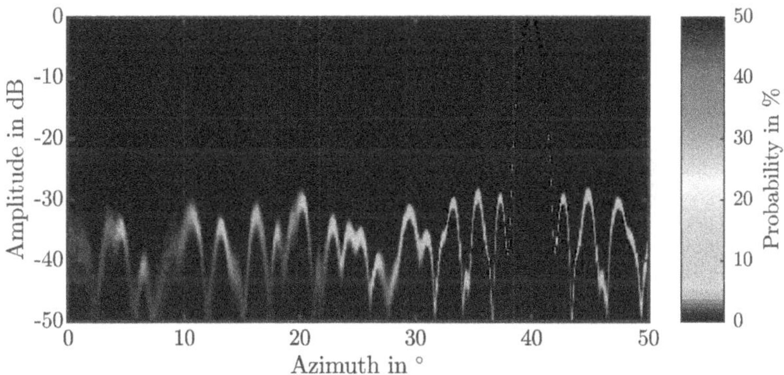

Figure 6.5: Effect of known random module displacement of up to ±675 µm for a single target from 40 ° azimuth, and zero elevation at 79 GHz. The probability is limited to 50 %.

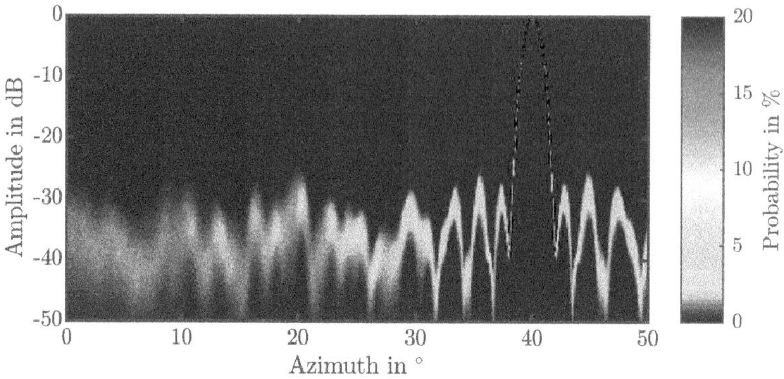

Figure 6.6: Effect of known random module displacement of up to ±491 µm for a single target from 40° azimuth, and zero elevation at 240 GHz. The probability is limited to 20 %.

As the previous simulations showed, the impact of manufacturing tolerances and thermal expansion alone is enough to significantly degrade the system's performance. This manifests as an increased SLL, especially close to a target. The main lobe is also impacted in both amplitude and angle. The problem becomes much more significant at higher frequencies. Further investigation of the problem revealed that the knowledge of the actual antenna displacement helps to restore the main lobe completely and reduce the impact on the SLL significantly. Therefore, it is logical that a system calibration would increase the performance, as it resolves the issues almost completely.

6.3 Calibration Algorithm

As the displacement of the modules is not known directly, it has to be measured somehow. A standard method is to calibrate the radar after production or after installation. This would involve an additional production step, which has to be carried out after the car's complete assembly and is thus not preferred. Additionally, a factory calibration will not compensate for displacements due to thermal expansion or aging processes. A calibration running on-the-fly during regular operation would be an ideal solution. Such a calibration algorithm is proposed in the following.

The proposed calibration process is based on the fact that the phase values for all transmitter to receiver combinations for an individual target only depend on the target's angles (azimuth and elevation) and maybe on the target's distance. If the beamforming factors are applied to each transmit/receive combination without summing them up, all phase values should be the same in an ideal situation. It is then possible to calculate a correction value for the module positions and slowly optimize the positions within an iterative process based on multiple targets. As the modules' displacement does not change rapidly, the system has a lot of time to collect potential targets for calibration.

First, the estimated displacement values $\boldsymbol{E}_{o,g}$ are set to zero for the first iteration $o = 0$ and all modules g:

$$\boldsymbol{E}_{o,g}\big|_{o=0} = \begin{pmatrix} 0 \\ 0 \\ 0 \end{pmatrix} . \tag{6.1}$$

The radar system then performs beamforming, which is done anyway during regular operation. Particularly clear reflections with a high SNR are then detected with a second CFAR algorithm. The CFAR can also be used to make sure that the reflection is likely to be just from a single target and not multiple ones by reducing the guard cells. One may even add an additional threshold for the training cells to prevent the detection of reference targets with other targets close by. This is important for the correct operation of the calibration algorithm, as will be explained later.

The next step is to determine the azimuth and elevation angles of that reference target with a high resolution. The resolution should be much higher than the system's angular resolution. This is possible if a single target is assumed to cause the reflection. Obviously, this assumption will not hold for every detected target, but the algorithm's iterative nature will average out these errors. The high angular resolution estimation implementation is trivial, with a high-resolution grid around the target and normal beamforming. Other peak search algorithms, like Newton's method, may be used to increase the speed of the algorithm. The estimation of the target's angle is based on the antenna positions, which do not include the displacement and is thus susceptible to errors. However, the angle estimation is good enough to let the system slowly improve itself, as will be shown later.

Once the target's direction is known, the originally received signals are processed according to the modulation type used in the system, and the samples for the chosen target are extracted.[1] The result is one complex number $g_{\delta,f_\mathrm{d},t,r}$ for each transmit/receive combination. The delay δ and the Doppler frequency f_d are assumed to be constant for all transmit/receive combinations, which should apply for most systems, even in near field conditions, as discussed in Section 4.2. These samples are then multiplied with the corresponding beamforming factor, which includes the currently estimated displacement, according to

$$h_{t,r} = g_{\delta,f_\mathrm{d},t,r} \left(\boldsymbol{B}^{\mathrm{ff}}\right)_{t,r} , \tag{6.2}$$

resulting in the corrected receive samples $h_{t,r}$. This processing only has to be carried out for the distance/delay δ and Doppler f_d the target is located at, and those indices are omitted subsequently.

If the estimated displacement is correct, the phase values of all corrected values should be the same after the last operation, as this is one goal of beamforming: Manipulate the phase values of each transmit/receive combination in such a way that the added signals sum up constructively (i.e., with the same phase) when pointing towards the target.

As the next step, it is necessary to subtract the average phase with

$$h_{\mathrm{norm},t,r} = \frac{h_{t,r}}{\frac{1}{N_\mathrm{t} N_\mathrm{r}} \sum_{t=1}^{N_\mathrm{t}} \sum_{r=1}^{N_\mathrm{r}} h_{t,r}} , \tag{6.3}$$

[1] For example, in a CS-based system, the range-Doppler matrix for each transmit/receive combination is calculated, and the targets are detected. Each target is then processed individually by extracting the scalar amplitudes for each transmitter/receiver combination from the corresponding range-Doppler matrix.

which is the division by the mean value over t and r. This is required because the absolute phase may be a more or less random value, depending on the exact position of the target.

As individual modules are expected to be displaced as a group, the average offset between two modules is calculated by using all transmitters from one module and all receivers from another module, resulting in the value

$$h_{\mathrm{mod},g_\mathrm{t},g_\mathrm{r}} = \sum_{t \text{ in } g_\mathrm{t}} \sum_{r \text{ in } g_\mathrm{r}} h_{\mathrm{norm},t,r}, \tag{6.4}$$

where g_t and g_r are the module indexes containing the corresponding transmitter and receiver antenna indexes, respectively. For example the module $g_t = 1$ contains the transmitters $t = 1..6$ and the module $g_r = 2$ contains the receivers $r = 9..16$, which is depicted in Table 8.1 and Table 8.2 for the transmitters and receivers, respectively. (6.4) then sums the samples of all combinations of t and r for the given groups. This is equivalent to a beamforming process, which includes only the transmitters from g_t and the receivers from g_r. At this point, it is also obvious why the target should be very clear and have no other targets close by. The aperture of this beamforming process is much smaller than the full array, and a much wider beam has to be expected. It is important that only the selected target is primarily visible by that beam.

The phase of the resulting offset value can be converted into an equivalent distance offset according to

$$\Delta d_{g_\mathrm{t},g_\mathrm{r}} = \frac{1}{k}\angle h_{\mathrm{mod},g_\mathrm{t},g_\mathrm{r}} = \frac{\lambda}{2\pi}\angle h_{\mathrm{mod},g_\mathrm{t},g_\mathrm{r}}. \tag{6.5}$$

An illustration for the calculation of the displacement distance is shown in Figure 6.7. In this case, the transmitter and receiver are too far away from each other, resulting in a phase/delay offset between the expected ideal position and the actual measured value.

The measured distance error only contains information on the actual displacement in the direction of the target. If the target's direction is orthogonal to the displacement, the measured phase/distance offset will be zero. As the antennas are all placed in the yz-plane, targets in boresight direction contain no displacement information as they are always orthogonal to any displacement. Therefore, targets with an absolute azimuth or elevation value less than 20° are not used for the calibration process.

The correction value is calculated with

$$\Delta \boldsymbol{E}_{g_\mathrm{t},g_\mathrm{r}} = \beta \Delta d_{g_\mathrm{t},g_\mathrm{r}} \begin{pmatrix} 0 \\ -\cos(\delta_\mathrm{tgt})\sin(\gamma_\mathrm{tgt}) \\ \sin(\delta_\mathrm{tgt}) \end{pmatrix}, \tag{6.6}$$

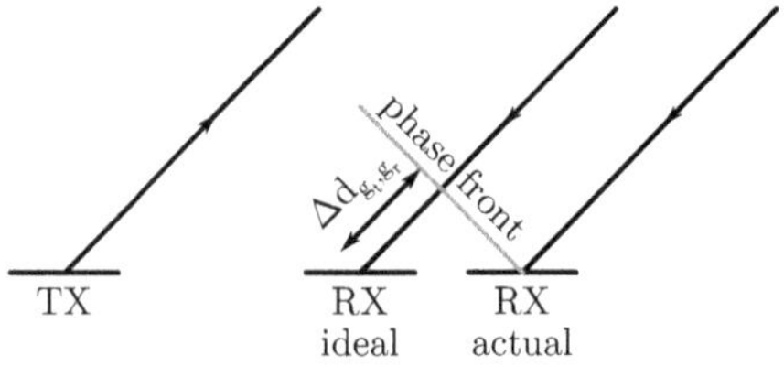

Figure 6.7: Illustration of an example for antenna displacement

where β is a scaling factor used to control the convergence speed of the algorithm. The vector in (6.6) is basically $\boldsymbol{p}_{\theta,\varphi}$ from (2.39) on page 19, but the value for the x-axis is set to zero, as all antennas are located on the yz-plane, and no displacement in the x-axis is assumed. δ_{tgt} and γ_{tgt} are the elevation and azimuth values of the target, respectively. The vector contains the y and z components pointing towards the target. This is required as a single target only contains offset information in the target's direction. If, for example, a target is located in the $z = 0$ plane, an offset of the antennas in the z-axis would have no impact on the measured offset distance $\Delta d_{g_\text{t},g_\text{r}}$. In this case, only corrections in the y-axis will be applied.

The correction value is then added to the two displacement vectors of the modules g_t and g_r for the next iteration $o+1$:

$$\boldsymbol{E}_{o+1,g_t} = \boldsymbol{E}_{o,g_t} + \Delta\boldsymbol{E}_{g_\text{t},g_\text{r}}, \tag{6.7}$$

$$\boldsymbol{E}_{o+1,g_r} = \boldsymbol{E}_{o,g_r} + \Delta\boldsymbol{E}_{g_\text{t},g_\text{r}}. \tag{6.8}$$

The process then starts from the beginning with the updated displacement vectors and a new target.

To get a scalar measure for the overall displacement error, the root mean square (RMS) offset of all modules to their actual position is used. Of course, this value can only be calculated if the actual positions are known, and therefore simulations are used to evaluate the algorithm. Before the RMS offset can be calculated, an additional step is required: The mean module offset in both axes has to be subtracted from the estimated module offset. This is required, as the average estimated offset may drift in a random direction. This drift is not noticed by the algorithm, especially in far-field conditions, because the beamforming result does not change with a constant offset in all antenna positions. The correction is carried out by removing the mean value of all displacement vectors from the current displacement vector:

$$\boldsymbol{E}_{\text{nm},o+1,g} = \boldsymbol{E}_{o+1,g} - \text{mean}_g \left\{\boldsymbol{E}_{o+1,g}\right\}. \tag{6.9}$$

The RMS error in the module positions is then calculated with

$$\epsilon_{o+1} = \sqrt{\frac{1}{N_\text{g}}\sum_{g=1}^{N_\text{g}} \left(\boldsymbol{E}_{\text{nm},o+1,g,\text{y}} - \boldsymbol{E}_{\text{i},g,\text{y}}\right)^2 + \frac{1}{N_\text{g}}\sum_{g=1}^{N_\text{g}} \left(\boldsymbol{E}_{\text{nm},o+1,g,\text{z}} - \boldsymbol{E}_{\text{i},g,\text{z}}\right)^2}, \tag{6.10}$$

with N_g being the number of modules in the system and $\boldsymbol{E}_{\text{i},g}$ are the actual module offsets.

Although the described algorithm is using far-field beamforming in (6.2), it can easily be adopted to near-field beamforming by using (2.30) instead. Everything else works equivalently.

6.4 Simulation

Figure 6.8 shows the displacement error over 2000 iterations for an exemplary scenario. It uses the same setup as before: A random displacement of up to ±675 µm is applied to the y and z axes of the proposed antenna array modules from Chapter 3 at a frequency of 79 GHz. The actual displacement values are listed in Table 6.2. The randomly generated targets have an SNR of either 15 dB or 30 dB after beamforming with all antennas, which

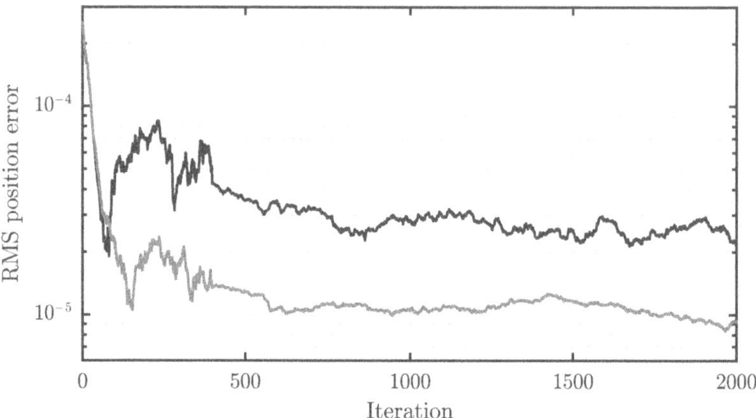

Figure 6.8: Example of the remaining RMS displacement of the module position optimization with a maximum random displacement of up to ±675 µm in y and z axes at the beginning and an SNR of 15 dB (blue) and 30 dB (red)

Table 6.2: Displacement values used for simulation shown in Figure 6.8

Module	Displacement y	Displacement z
1	-521.8 µm	278.1 µm
2	640.6 µm	-342.0 µm
3	308.8 µm	-329.4 µm
4	-200.5 µm	-642.6 µm
5	280.3 µm	-541.7 µm
6	404.5 µm	-269.4 µm
7	196.5 µm	190.2 µm
8	-115.3 µm	-240.0 µm

corresponds to the SNR seen during normal operation. It is obvious that the higher SNR is preferable as the final estimation for the displacement is much better.

As described in the previous section, the value of β is used as a factor for the correction vector $\Delta\boldsymbol{E}_{g_\mathrm{t},g_\mathrm{r}}$. It influences the speed of convergence, and the minimum value reached. In general, the measurements are noisy, and this noise will propagate through to the correction vector. A smaller value for β will limit the noise, as it acts similar to a low pass filter. However, the algorithm will require longer to reach a certain value. If the value is larger, more noise passes through to the currently estimated displacement vector, increasing the final position offset, but the algorithm will converge faster.

This behavior can be observed in Figure 6.8: β is initially set to a relatively high value of 10^{-2}, allowing a fast convergence within the first 100 iterations. After that, the algorithm reached its final value, but the estimated displacement contains a lot of noise. At iteration

400, the value of β is changed to 10^{-3}, immediately reducing the impact of the system noise and allowing the algorithm to reduce the RMS offset value even further, but slower than before.

By using the approach of a variable value for β the algorithm is capable of reducing the RMS offset from initially 256 µm down to approximately 20 µm within the first 100 iterations in case of targets with an SNR of 30 dB. This corresponds to more than a decade of improvement, and the final value is less than $\lambda/100$, which is sufficient for operation. After reducing the value of β, the displacement's error is stabilized and reduced to a value as small as approximately 10 µm.

6.5 Summary

This chapter analyzed potential impacts on the antenna positions during manufacturing and operation. The two major influences, namely the mechanical manufacturing tolerances and the thermal expansion, were investigated further and showed that those parameters alone might cause a worst-case displacement of up to 400 µm and 275 µm, respectively. Subsequently, the effect of a random displacement of the antenna modules proposed in Chapter 3 of up to 675 µm in total was analyzed at a frequency of 79 GHz. This simulation showed substantial degradation of the SLL and minor degradation of the main lobe in both amplitude and angle. The effect is much more significant for higher frequencies, shown with a simulation at 240 GHz.

A calibration algorithm was proposed to compensate for the effect of unknown antenna displacements mostly. The algorithm can run during the operation of the system and utilizes available targets in the scenario. It calculates the difference between expected and measured phase values and can estimate the actual antenna positions in an iterative process with multiple targets. The proposed algorithm's operation was shown to work in simulations as expected when targets with a high SNR are available. It reduced the RMS module displacement of initially 256 µm down to 10 µm, which is sufficient for proper operation of the radar.

7 Conclusions and Future Work

Future radar systems for assisted and autonomous driving will require a higher resolution than currently available automotive radar systems. Compared to other sensors like lidar or camera systems, radars lack angular resolution but are competitive in other areas like cost and robustness in harsh conditions. The two critical components for such a system are a high bandwidth, which results in a high range resolution, and a large aperture, which increases the angular resolution. This thesis established the required general framework for such a radar system and proved its feasibility with two demonstrators.

Chapter 3 laid the foundation by introducing all requirements to a future radar system. A modular approach seemed reasonable to reduce the complexity of the hardware. Two genetic algorithms were proposed and adapted to fit the needs of optimizing a large aperture MIMO antenna array. The algorithms optimized the antenna elements' positions and weights to achieve an excellent angular resolution while keeping the side lobe level as low as possible. The proposed array has an angular resolution of 1 ° in azimuth, which is better by a factor of approximately seven compared to currently available automotive radar systems. The array was thoroughly analyzed, and its performance within a typical radar system was estimated, proving the detection capability for common targets up to a distance of at least 150 m.

The first demonstrator, described in Chapter 4, showed the concept of a modular radar system and the coherent synchronization of the individual modules. The synchronization was carried out with a radio over fiber system, demonstrating the feasibility of optical systems combined with a modular radar system. The combination of optical and electronic components may become more relevant in future systems if the integration of both technologies into a single process becomes viable in terms of cost. The antenna array employed for the demonstrator provided a high angular resolution, but with limited side lobe level to reduce the system's complexity. The demonstrator showed the expected performance in an anechoic chamber and proved the general concept to be valid.

Chapter 5 described a second demonstrator, which implemented the large array proposed in Chapter 3. It used the virtual aperture and scanned it with a synthetic aperture radar. The system was built with an off-the-shelf radar module mounted on an XY-table. The demonstrator was used in a more realistic automotive outdoor scenario, including different targets like a car, a bicycle, or a building's wall. The off-line signal processing was able to reconstruct a three-dimensional image of the scenario, showing all targets with the expected resolution. The measurement showed the capabilities of the proposed array configuration from Chapter 3 as a viable approach for future automotive radar systems.

Finally, Chapter 6 discussed the causes of not precisely known antenna positions and the potential impact on the radar's performance at different frequencies. Several sources for tolerances were identified, like mechanical tolerances during production or thermal expansion, and a worst-case estimation for each of them was made. A statistical analysis of thousands of simulations showed an increased side lobe level and a reduced main lobe with substantial angle errors for radars operating at higher frequencies. Subsequently, an algorithm was proposed and tested in simulations to estimate the actual antenna positions in the array and compensate for most of the negative influences due to manufacturing tolerances, thermal expansion, and other sources. The proposed algorithm can run during the radar's regular operation and can compensate for effects occurring after production.

Although the demonstrators are based on a 79 GHz radar, future radar systems are pushing to higher frequencies. This allows the usage of a higher bandwidth and thus a better range resolution, and it keeps the required aperture size compact, as the angular resolution is proportional to the wavelength. However, the results presented in this work apply to any operation frequency of the radar. Another problem when introducing a large number of transmitters is the required time to scan through all of them, especially when Doppler information is required, and the maximum time between the scans of an individual transmitter is limited. A possible solution to this problem is to use a different modulation type, which can achieve orthogonal signals, even when transmitted simultaneously, like, for example, PN sequences. Additionally, further testing of the module displacement correction algorithm from Chapter 6 in real scenarios is required to prove its capabilities in real scenarios.

One of the significant next steps would be to integrate the proposed array into a completely functional prototype, which does not rely on a synthetic aperture. This would allow evaluating the system in even more realistic scenarios, including moving targets and a moving platform. An important step would also be to improve the overall signal processing in terms of efficiency. This would probably lead to an implementation in a faster language, like C or C++, and porting to more efficient hardware, like graphic cards, FPGAs, or even ASICs.

8 Appendices

8.A Antenna Element Positions

Table 8.1: Positions of the transmitters (t) of the proposed array and their corresponding module (M)

t	M	Y in $\lambda/2$	Z in $\lambda/2$	t	M	Y in $\lambda/2$	Z in $\lambda/2$
1	1	4	14	25	5	8	30
2	1	8	8	26	5	6	17
3	1	1	1	27	5	10	20
4	1	8	2	28	5	17	28
5	1	4	2	29	5	7	20
6	1	2	8	30	5	9	20
7	2	22	12	31	6	26	28
8	2	30	5	32	6	23	20
9	2	27	7	33	6	35	20
10	2	35	13	34	6	22	20
11	2	37	7	35	6	29	25
12	2	26	1	36	6	36	28
13	3	44	3	37	7	52	17
14	3	52	11	38	7	46	23
15	3	39	14	39	7	50	20
16	3	47	0	40	7	41	24
17	3	47	9	41	7	50	26
18	3	53	6	42	7	48	25
19	4	65	12	43	8	59	20
20	4	62	5	44	8	64	16
21	4	75	3	45	8	73	27
22	4	64	4	46	8	61	18
23	4	70	12	47	8	69	26
24	4	63	0	48	8	71	28

Table 8.2: Positions of the receivers (r) of the proposed array and their corresponding module (M)

r	M	Y in $\lambda/2$	Z in $\lambda/2$	r	M	Y in $\lambda/2$	Z in $\lambda/2$
1	1	15	13	33	5	7	17
2	1	10	12	34	5	3	23
3	1	6	12	35	5	9	16
4	1	15	9	36	5	1	26
5	1	17	6	37	5	7	21
6	1	9	1	38	5	7	16
7	1	2	1	39	5	9	27
8	1	14	14	40	5	15	23
9	2	24	7	41	6	22	19
10	2	23	6	42	6	29	30
11	2	30	2	43	6	32	24
12	2	35	2	44	6	27	25
13	2	35	11	45	6	24	17
14	2	27	13	46	6	24	26
15	2	35	1	47	6	31	16
16	2	37	1	48	6	33	23
17	3	48	11	49	7	56	17
18	3	46	5	50	7	41	17
19	3	43	1	51	7	54	26
20	3	48	9	52	7	47	20
21	3	49	3	53	7	42	29
22	3	45	7	54	7	41	20
23	3	39	1	55	7	55	18
24	3	43	2	56	7	40	20
25	4	69	11	57	8	64	22
26	4	63	13	58	8	68	27
27	4	70	10	59	8	68	17
28	4	72	3	60	8	74	20
29	4	75	14	61	8	74	25
30	4	59	11	62	8	62	20
31	4	75	12	63	8	62	24
32	4	59	8	64	8	72	27

8.B Antenna Element Weights

Table 8.3: Antenna weights for the transmitters and receivers from 1 to 16. The weight is given per mille (i.e., thousandth).

t \ r	1	2	3	4	5	6	7	8	9	10	11	12	13	14	15	16
1	548	538	459	326	536	432	529	699	411	468	483	718	334	374	395	385
2	321	585	475	401	552	528	391	667	273	410	414	518	505	537	517	374
3	673	346	538	698	597	407	324	514	443	675	150	424	337	446	451	411
4	579	409	497	800	289	441	455	333	355	749	633	522	385	544	83	363
5	523	574	324	324	360	584	498	675	526	591	370	385	724	421	367	495
6	349	443	516	534	444	557	488	416	610	488	395	269	213	715	540	670
7	456	629	588	426	590	637	409	357	853	458	515	629	831	560	698	483
8	573	703	618	654	452	617	244	566	504	665	576	434	330	488	802	505
9	657	238	586	789	362	572	420	346	543	481	530	427	550	572	489	290
10	531	733	475	484	544	601	850	613	529	580	419	440	505	561	884	498
11	720	570	597	687	285	573	535	721	421	650	626	596	365	545	385	436
12	819	587	583	389	479	582	592	691	803	382	417	612	236	634	690	453
13	483	538	429	644	709	282	437	502	333	347	479	460	347	431	309	573
14	590	427	443	554	375	252	435	555	491	763	610	693	258	444	402	567
15	633	496	350	336	603	622	477	613	403	803	682	568	408	559	544	286
16	454	704	596	510	293	559	521	415	540	557	683	423	577	330	479	277
17	327	450	374	417	700	376	435	685	404	437	672	507	490	525	616	639
18	658	635	550	567	232	486	617	648	392	597	584	681	388	607	633	599
19	602	674	700	654	464	272	552	287	659	806	761	201	600	448	691	372
20	563	565	365	189	452	467	365	549	458	586	278	386	528	588	585	327
21	395	722	743	625	457	382	406	392	93	568	430	663	569	564	330	310
22	513	604	393	483	528	683	638	559	531	551	627	439	694	538	403	536
23	273	744	425	491	463	353	738	548	495	460	361	525	465	548	518	554
24	654	412	587	536	617	487	461	331	543	304	348	252	648	782	353	468
25	556	390	447	433	374	310	533	238	266	507	431	845	705	556	604	310
26	672	200	534	568	584	260	399	407	460	467	898	630	487	511	486	661
27	576	486	419	561	341	634	357	324	658	526	544	404	421	349	623	625
28	483	550	546	312	650	578	473	256	495	554	918	343	732	676	280	594
29	682	241	487	351	555	430	339	581	394	654	158	578	486	713	429	320
30	608	435	318	466	545	448	387	484	472	706	366	566	703	688	445	511
31	563	612	429	492	588	478	423	320	456	466	620	622	457	583	335	416
32	250	435	561	561	506	635	368	509	550	608	658	400	465	420	456	522
33	359	514	677	485	844	333	342	628	221	305	601	417	762	822	445	670
34	448	675	748	535	470	521	466	190	506	298	347	531	460	617	528	427
35	624	500	526	249	396	504	475	528	757	692	346	619	926	475	536	464
36	660	467	594	576	830	360	679	455	819	425	403	469	891	720	310	751
37	453	494	308	448	552	707	470	474	372	748	617	975	584	375	498	650
38	684	440	539	666	319	777	513	619	400	528	461	668	555	782	419	350
39	914	830	608	448	483	499	338	703	376	494	725	369	528	568	620	337
40	502	510	675	619	704	423	411	703	397	667	313	557	440	523	447	517
41	537	740	453	639	529	393	269	502	587	623	651	856	645	397	557	755
42	695	482	452	692	286	452	264	723	638	291	236	291	417	388	512	607
43	497	560	760	699	518	670	508	672	488	586	816	708	472	627	792	813
44	583	363	518	1000	691	705	726	601	761	238	374	603	348	514	665	268
45	644	553	351	493	638	432	568	525	446	702	656	360	425	534	367	333
46	368	643	464	306	304	532	605	500	521	377	445	710	444	514	401	335
47	625	438	330	446	351	429	714	393	634	530	418	747	192	591	557	372
48	776	390	558	512	523	302	695	664	635	668	547	570	725	439	514	377

Table 8.4: Antenna weights for the transmitters and receivers from 17 to 32. The weight is given per mille (i.e., thousandth).

t \ r	17	18	19	20	21	22	23	24	25	26	27	28	29	30	31	32
1	573	411	477	666	545	434	680	461	704	369	709	613	637	559	504	558
2	572	390	515	648	506	630	418	408	566	424	479	602	524	484	748	707
3	516	302	515	380	401	561	427	298	430	642	715	384	452	552	323	613
4	433	600	565	460	424	549	298	235	275	429	491	376	408	470	707	469
5	266	577	159	667	545	409	142	505	516	430	310	421	551	583	375	360
6	519	416	477	682	368	453	545	566	412	362	482	406	694	742	495	489
7	597	487	360	585	431	450	302	495	507	393	692	578	568	408	303	731
8	412	544	471	435	741	749	323	519	492	343	580	602	657	470	480	504
9	690	161	413	510	413	466	614	251	523	678	730	233	650	911	408	679
10	410	586	591	387	592	701	545	370	639	705	626	440	762	755	677	782
11	811	806	501	602	458	524	808	357	323	206	457	365	626	441	283	486
12	352	649	463	760	593	300	521	305	316	933	803	506	522	528	488	234
13	676	415	456	692	578	426	749	627	51	398	432	365	226	726	287	708
14	573	520	306	282	355	633	500	730	422	552	300	423	554	391	507	649
15	442	306	735	728	295	676	341	656	341	653	413	282	610	654	487	302
16	719	413	277	341	521	313	287	516	575	473	576	583	399	327	243	612
17	573	773	645	609	723	586	246	417	376	540	481	566	160	819	449	670
18	399	677	591	317	481	464	316	635	632	677	416	604	512	600	946	507
19	810	144	560	496	324	545	690	560	494	550	623	554	629	432	641	307
20	464	477	337	689	536	525	596	386	861	720	593	494	637	805	304	628
21	806	678	495	350	513	544	582	675	547	359	117	467	289	283	377	384
22	623	450	459	391	383	377	533	436	522	418	147	24	412	677	528	367
23	488	344	278	533	223	642	585	226	515	438	638	624	228	496	672	364
24	507	500	502	449	403	500	242	533	595	517	415	331	383	228	274	308
25	855	792	566	636	598	347	589	504	512	594	478	540	590	420	589	744
26	360	428	614	313	713	375	303	691	567	539	302	571	527	458	689	626
27	428	381	488	418	588	461	319	393	580	498	697	918	587	491	662	483
28	525	387	630	560	295	706	377	686	526	633	612	492	591	407	695	628
29	235	718	503	894	570	489	447	500	484	635	649	518	752	430	577	572
30	521	525	430	659	541	430	484	393	736	442	328	363	772	654	348	542
31	388	376	349	433	561	454	566	438	581	713	537	586	614	627	548	614
32	434	844	589	331	549	652	530	372	292	636	533	384	622	771	499	427
33	421	674	546	785	632	522	655	654	514	518	641	390	572	363	509	385
34	727	683	404	832	416	442	571	604	698	697	469	500	305	571	576	563
35	643	610	426	740	578	587	733	412	674	460	436	283	607	600	264	583
36	555	434	373	543	400	711	536	414	430	531	430	682	476	657	406	689
37	557	337	621	370	744	264	588	755	541	284	380	367	259	780	399	556
38	597	329	569	601	732	444	558	528	654	531	572	382	472	148	770	639
39	353	575	592	463	516	733	727	628	364	511	464	61	665	566	608	648
40	783	559	683	642	854	668	623	445	375	349	509	485	515	745	340	869
41	706	457	739	263	419	538	346	670	391	328	399	524	628	613	455	417
42	714	759	689	552	756	636	493	612	750	722	391	499	461	552	736	482
43	471	638	482	495	569	521	429	497	363	277	499	699	324	473	458	399
44	267	491	277	578	471	272	336	654	450	412	601	638	456	589	468	512
45	711	583	510	752	606	369	396	547	492	508	527	489	248	518	427	350
46	412	644	590	861	496	372	553	417	435	509	563	287	322	657	319	657
47	608	585	324	368	579	620	626	452	534	446	527	208	211	736	362	569
48	530	494	386	525	460	614	509	663	399	290	502	591	629	458	403	342

Table 8.5: Antenna weights for the transmitters and receivers from 33 to 48. The weight is given per mille (i.e., thousandth).

t \ r	33	34	35	36	37	38	39	40	41	42	43	44	45	46	47	48
1	283	437	621	420	496	612	644	478	410	395	612	211	414	466	534	707
2	613	272	598	326	801	501	531	557	368	661	675	758	444	488	547	823
3	313	390	382	405	722	615	309	403	545	529	511	596	509	621	685	186
4	341	410	517	530	430	359	268	341	304	416	620	243	504	593	269	450
5	498	339	550	365	452	514	312	615	383	379	282	341	396	668	410	372
6	221	421	578	368	224	511	379	221	435	643	452	450	625	193	543	516
7	529	386	466	440	462	414	371	536	724	527	734	777	457	797	605	538
8	272	523	306	551	594	452	618	554	595	681	592	568	754	525	370	741
9	287	468	641	506	761	652	231	308	359	453	585	400	369	627	335	635
10	317	574	661	453	693	565	504	644	774	804	729	691	528	491	727	290
11	346	576	361	255	482	787	631	367	392	204	444	416	369	300	545	360
12	436	578	332	537	510	716	739	593	287	358	430	672	526	596	527	514
13	538	582	738	320	416	375	798	627	386	429	270	674	615	574	686	471
14	739	555	441	392	510	359	591	663	616	518	453	515	450	577	470	344
15	476	682	568	468	796	571	445	456	353	567	509	585	594	464	254	316
16	524	435	528	709	527	537	841	796	596	367	321	375	746	506	489	598
17	242	630	510	677	788	596	585	506	568	324	544	558	538	423	530	707
18	852	220	450	521	758	311	437	380	535	826	600	412	437	574	708	676
19	512	397	211	409	495	921	444	408	684	602	697	324	279	355	510	310
20	457	444	359	778	558	619	596	358	557	925	451	423	562	623	692	802
21	514	680	594	384	583	542	571	444	728	544	785	541	474	345	694	590
22	437	323	609	539	440	688	455	649	791	558	636	443	569	547	574	723
23	343	549	420	367	402	539	542	497	581	405	535	322	295	685	425	545
24	222	555	400	359	446	307	344	645	672	429	468	769	595	646	463	479
25	427	481	415	339	435	435	501	459	209	212	419	233	407	423	311	581
26	419	178	421	553	728	520	712	559	757	471	466	511	883	339	325	566
27	413	231	658	350	537	464	645	378	678	579	464	631	350	636	729	366
28	199	410	488	554	627	391	647	224	308	331	521	535	520	481	410	500
29	617	654	484	321	528	466	498	556	497	611	348	754	642	233	542	375
30	498	455	531	423	702	306	303	418	294	586	614	464	649	603	792	570
31	281	536	609	669	545	432	367	452	395	291	672	459	561	350	537	718
32	419	339	318	139	460	442	323	242	749	541	504	486	656	489	523	529
33	531	361	351	336	722	630	290	538	604	572	478	529	633	851	579	264
34	615	440	312	688	684	432	496	489	364	395	489	509	340	544	643	551
35	423	430	465	436	394	563	619	472	437	642	310	780	570	596	754	495
36	500	525	421	376	609	491	334	372	503	606	588	502	312	422	646	615
37	273	661	491	297	496	498	653	681	772	425	817	717	583	297	384	447
38	553	662	599	531	435	621	254	434	562	512	319	193	493	427	607	443
39	585	399	599	496	442	610	316	266	427	458	408	374	198	549	639	508
40	412	249	301	391	434	597	410	569	576	346	260	398	489	363	298	453
41	503	308	725	412	700	427	567	352	396	342	345	431	514	416	619	675
42	565	415	477	555	546	732	388	426	390	518	837	510	541	624	859	431
43	347	712	169	545	716	624	536	410	543	424	709	496	423	519	837	647
44	937	782	512	565	397	588	452	451	585	256	323	522	723	667	671	785
45	302	519	711	535	485	554	404	351	387	376	432	338	409	311	422	502
46	552	377	382	452	622	667	532	413	342	525	445	666	551	214	429	635
47	422	654	378	324	558	597	734	582	259	386	475	395	443	374	741	428
48	565	345	424	595	698	544	528	510	313	385	291	447	464	628	507	506

Table 8.6: Antenna weights for the transmitters and receivers from 49 to 64. The weight is given per mille (i.e., thousandth).

t \ r	49	50	51	52	53	54	55	56	57	58	59	60	61	62	63	64
1	883	480	475	514	498	665	614	470	425	829	872	553	407	546	560	517
2	558	543	531	450	352	328	472	376	476	348	195	469	380	519	572	281
3	383	441	511	432	660	584	341	399	614	376	366	285	467	351	540	627
4	715	456	757	598	476	521	542	231	815	212	403	692	657	509	455	459
5	552	517	654	552	381	424	407	542	346	578	535	526	233	317	490	498
6	406	306	462	377	671	618	656	430	523	832	702	534	433	655	536	590
7	649	167	651	540	450	342	593	558	348	695	527	717	508	403	636	467
8	609	727	554	604	520	696	693	452	554	310	336	442	330	401	522	692
9	348	386	532	473	534	467	743	321	388	453	424	500	728	375	461	667
10	572	497	508	387	517	414	756	791	508	552	561	494	656	802	487	913
11	517	223	781	976	578	191	424	774	325	368	578	303	606	200	497	349
12	393	727	438	685	616	481	563	660	601	504	362	400	377	512	804	488
13	248	589	482	581	487	663	632	321	381	693	429	493	298	413	569	412
14	336	611	492	565	404	612	673	323	508	661	526	715	465	720	571	314
15	657	314	407	175	577	430	492	622	576	526	337	551	571	627	453	464
16	270	346	516	636	339	587	503	567	533	464	513	569	632	493	375	806
17	479	638	786	359	371	440	559	665	584	558	488	441	507	519	437	693
18	149	604	454	403	520	531	562	760	453	717	482	469	195	427	332	476
19	375	315	409	687	476	556	475	319	516	591	546	297	593	518	353	194
20	487	518	469	513	590	350	545	727	728	605	508	676	327	179	376	346
21	292	427	549	128	272	413	618	454	509	570	394	734	526	303	289	438
22	650	678	510	512	449	577	433	516	400	257	590	467	366	565	491	379
23	421	578	459	345	246	278	626	688	393	373	254	346	490	153	400	360
24	264	717	474	472	243	470	564	727	508	508	443	396	406	412	659	624
25	396	700	410	273	348	684	323	554	429	530	459	445	490	573	435	320
26	426	624	634	300	499	491	895	500	567	486	457	798	653	429	510	369
27	315	569	372	386	376	633	469	173	611	376	292	466	462	463	477	495
28	443	539	634	558	761	388	494	606	386	558	216	631	617	428	521	221
29	451	482	432	506	481	629	611	734	732	397	695	447	457	660	409	298
30	496	356	759	642	449	295	489	571	880	450	408	434	469	503	349	400
31	632	586	408	177	295	392	406	308	678	297	630	733	525	339	466	378
32	589	726	300	597	327	333	612	460	644	444	325	328	617	405	403	553
33	267	605	579	523	441	479	596	561	477	339	550	638	437	272	377	487
34	256	552	354	544	682	396	569	691	604	678	505	502	624	460	686	493
35	541	651	480	651	566	365	613	432	373	369	669	466	289	370	512	148
36	649	407	533	363	536	336	579	553	264	449	662	308	420	274	634	494
37	618	468	109	404	413	608	510	520	728	579	536	450	559	336	618	96
38	406	810	276	427	484	532	541	605	415	232	575	709	437	502	363	498
39	665	327	770	157	631	820	632	394	787	415	459	743	332	510	581	546
40	671	600	585	678	563	370	502	533	608	581	544	282	575	501	359	581
41	233	525	551	542	553	528	341	600	582	448	336	602	461	513	425	412
42	662	728	687	499	311	728	591	840	601	471	211	171	600	373	328	440
43	600	468	559	603	520	453	527	549	324	615	462	500	284	592	538	303
44	354	623	502	548	700	478	451	782	722	764	475	465	608	286	676	573
45	700	506	924	396	406	572	307	428	473	294	475	501	432	649	702	489
46	361	676	263	424	364	445	434	486	352	361	404	451	402	196	379	297
47	803	476	555	592	530	568	332	531	156	329	532	494	443	299	402	217
48	110	321	491	503	710	529	637	393	726	163	335	604	395	50	361	429

8.C Synchronization Demonstrator Components

The following list shows all major components, their manufacturer and part number used within the synchronization demonstrator, which was described in Chapter 4.

Table 8.7: Components of the synchronization demonstrator

Manufacturer	Part number	Description
Texas Instruments	AWR1243P	RoC - radar front end
Xilinx	XC7A50T-2FGG484C	FPGA
ISSI	IS25WP064A-JKLE	FPGA flash, 64 Mbit
FTDI	FT232HQ	High-speed USB to parallel bride
Atmel	ATSAM4S2AA-AU	Microcontroller
Sk Hynix	H5TC4G63EFR-PBA	DDR-3 memory, 4 Gbit
Samtec	QTH-030-01-X-D-A	High-speed connector (FPGA board)
Samtec	QSH-030-01-X-D	High-speed connector (radar board)
Optilab	PD20	20 GHz PD
Minicircuits	ZX60-183A-S+	20 GHz amplifier after PD
Analog Devices	HMC451LC3	20 GHz amplifier (radar board)
Oplink Comm.	ITMSE05ECO81111G	95/5 spllitter & Detector PD
Optilab	CQF975/208-19370	LD, 1547.71 nm, 10 mW
Optilab	IM-1550-20-TQ	20 GHz MZM, intensity mod.
LiComm	OFA-TCU-17AP-FA	Optical amplifier
GPON EPON	Unknown	Optical splitter
Thorlabs	TWQ1550HA	Optical splitter (alternative)
Thorlabs	MLD203CLN	LDC
Thorlabs	MTD415TE	TEC

8.D HPBW of a Corner Reflector

The equation for the RCS of a trihedral corner reflector is commonly known in literature, e.g. [19, p. 14.10][1], but the HPBW of such a reflector is less known. There are possibly may ways to calculate or estimate the HPBW of a trihedral corner reflector and two ways are illustrated below.

8.D.1 Analytical Comparision to a Plate

Unlike the HPBW of a trihedral corner reflector, the HPBW for a flat conducting plate is known, e.g. in [72, p. 401], as

$$\mathrm{HPBW}_{\mathrm{pl}} \approx 50.8\,^{\circ}\frac{\lambda}{a_{\mathrm{pl}}}, \tag{8.1}$$

which is, not surprisingly, the same as the HPBW for an aperture antenna, given in (2.24) on page 14. In this case a_{pl} denotes the dimension of the plate in the plane of interest.

The idea is to compare the two RCS values for both types of reflectors and come up with a scaling factor between the dimensions of a corner reflector (a_{cr}) versus the dimensions of a plate (a_{pl}), expecting a similar shaped main lobe in both cases. The two RCS values are defined as

$$\sigma_{\mathrm{pl}} = \frac{4\pi a^4}{\lambda^2}, \tag{8.2}$$

and

$$\sigma_{\mathrm{cr}} = \frac{4\pi a^4}{3\lambda^2}, \tag{8.3}$$

for the plate and the corner reflector, respectively [19, p. 14.10]. This means that a plate reflector needs to have a size of

$$a_{\mathrm{pl}} = \frac{1}{\sqrt[4]{3}} a_{\mathrm{cr}} \approx 0.76\, a_{\mathrm{cr}} \tag{8.4}$$

to match a corner reflector with a size of a_{cr}. Inserting this result into (8.1) yields an approximation for the HPBW of a corner reflector:

$$\mathrm{HPBW}_{\mathrm{cr}} \approx 66.9\,^{\circ}\frac{\lambda}{a_{\mathrm{cr}}}. \tag{8.5}$$

8.D.2 Simulation and Fitting

The second approach simulates the trihedral corner reflector with a multitude of size and wavelength combinations. The function

$$\mathrm{HPBW}_{\mathrm{cr}}(\Theta, \lambda, a_{\mathrm{cr}}) = \Theta\frac{\lambda}{a_{\mathrm{cr}}} \tag{8.6}$$

[1] Defined in the second paragraph of this page. The definition in table 14.1 refers to a dihedral corner reflector.

is then optimized to match the results as good as possible, where Θ represents the unknown scaling factor.

The simulations are carried out with Ansys Electronics Desktop using the Integral-Equation-Solver, which yields good results for electrically large problems while keeping the amount of required resources fairly low. The corner reflector is modelled with three triangular sheets of perfectly electric conductor (PEC). A plane wave excites the model and the far field of the reflected waves is evaluated. The simulation is carried out for frequencies between 50 and 150 GHz in 10 GHz steps and sizes between 20 and 80 mm in 10 mm steps. The far field is then further processed to calculate the HPBW for each frequency/size combination. The solid lines in Figure 8.1 depict the results of the simulation after calculating the HPBW.

The optimization of Θ in (8.6) is carried out with a simple optimization algorithm, which minimizes the overall relative RMS error between simulation and the equation. The error is defined as

$$\mathrm{err} = \sqrt{\sum_{\lambda, a_{\mathrm{cr}}} \left[\frac{\mathrm{HPBW}_{\mathrm{cr}}(\Theta, \lambda, a_{\mathrm{cr}}) - \mathrm{HPBW}_{\mathrm{cr,sim}}(\lambda, a_{\mathrm{cr}})}{\mathrm{HPBW}_{\mathrm{cr,sim}}(\lambda, a_{\mathrm{cr}})} \right]^2}, \tag{8.7}$$

where $\mathrm{HPBW}_{\mathrm{cr,sim}}(\lambda, a_{\mathrm{cr}})$ is the simulated HPBW for a given wavelength λ and size a_{cr}. The algorithm starts with an initial value for Θ, and calculates the error value according to (8.7) for Θ, $\Theta - \Delta_\Theta$, and $\Theta + \Delta_\Theta$, where Δ_Θ has a value of $0.01\,°$. The new value of Θ is the one which yielded the lowest error. This process is repeated until no better value is found.

Applying this algorithm to the simulated data from Figure 8.1 yields an optimal value of $65.9\,°$, resulting in the final equation for the HPBW of a corner reflector:

$$\mathrm{HPBW}_{\mathrm{cr}} \approx 65.9\,° \frac{\lambda}{a_{\mathrm{cr}}}. \tag{8.8}$$

The value is very close to the one calculated in (8.5), confirming the previous result.

The simulated HPBW in Figure 8.1 fluctuates around the average curve, represented by (8.8). This is expected, as the simulation includes certain effects, like resonances, which are not modelled by the final equation.

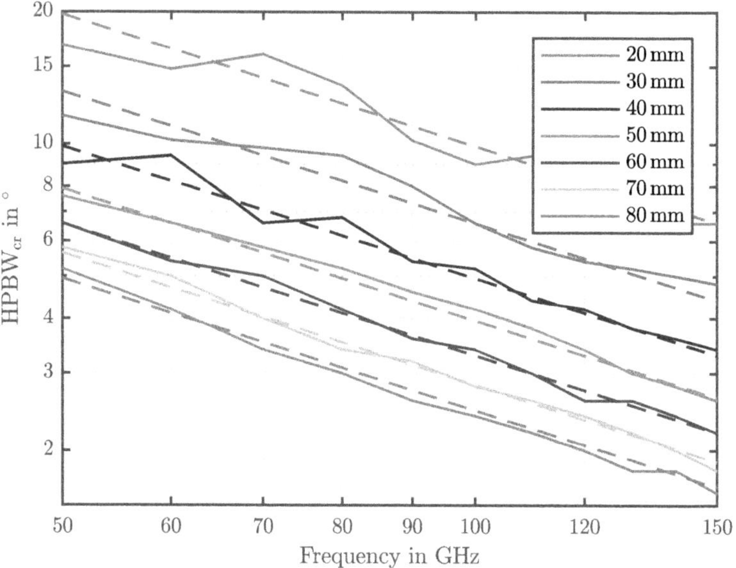

Figure 8.1: HPBW values for simulated corner reflector (solid) and the matched equation (dashed) in dependence of the frequency for different values of a_{cr}

List of References

[1] ITU, *Radio Regulations Articles.* International Telecommunication Union, 2012.

[2] L. Brown, *A Radar History of World War II: Technical and Military Imperatives.* Institute of Physics Publishing, 1999.

[3] J. Li and P. Stoica, *MIMO Radar Signal Processing.* Wiley, 2008.

[4] R. J. Mailloux, *Phased Array Antenna Handbook*, 3rd ed. Artech House, 2018.

[5] H. H. Meinel, "Evolving automotive radar – From the very beginnings into the future," in *The 8th European Conference on Antennas and Propagation (EuCAP 2014)*, 2014, pp. 3107–3114.

[6] J. Steinbaeck, C. Steger, G. Holweg, and N. Druml, "Next generation radar sensors in automotive sensor fusion systems," in *2017 Sensor Data Fusion: Trends, Solutions, Applications (SDF)*, 2017, pp. 1–6.

[7] A. Alaqeel, A. Ibrahim, A. Nashashibi, H. Shaman, and K. Sarabandi, "A Phenomenological Study of Radar Backscatter Response of Vehicles for the Next Generation Automotive Radars," in *IGARSS 2018 - 2018 IEEE International Geoscience and Remote Sensing Symposium*, 2018, pp. 4063–4065.

[8] K. Geary, J. S. Colburn, A. Bekaryan, S. Zeng, B. Litkouhi, and M. Murad, "Automotive radar target characterization from 22 to 29 GHz and 76 to 81 GHz," in *2013 IEEE Radar Conference (RadarCon13)*, 2013, pp. 1–6.

[9] M. Narandžić, "Wideband MIMO Radio Channel Modeling," dissertation, 2015. [Online]. Available: https://www.db-thueringen.de/servlets/MCRFileNodeServlet/dbt_derivate_00031817/ilm1-2015000112.pdf

[10] J. Gamba, *Radar Signal Processing for Autonomous Driving.* Springer, 2020.

[11] M. Jankiraman, *Design of Multi-Frequency CW Radars.* SciTech, 2007.

[12] C. A. Balanis, *Antenna Theory: Analysis and Design*, 4th ed. Wiley, 2016.

[13] H. J. Visser, *Antenna Theory and Applications.* Wiley, 2012.

[14] H. L. van Trees, *Optimum Array Processing: Part IV*, ser. Detection, Estimation, and Modulation Theory. Wiley, 2004.

[15] F. Roos, D. Ellenrieder, N. Appenrodt, J. Dickmann, and C. Waldschmidt, "Range migration compensation for chirp-sequence based radar," in *2016 German Microwave Conference (GeMiC)*, Mar. 2016.

[16] P. V. Brennan, S. Rahman, and L. B. Lok, "Range migration compensation in static digital-beamforming-on-receive radar," *IET Radar, Sonar & Navigation*, vol. 9, no. 9, pp. 1323–1329, 2015.

[17] A. D. Poularikas, *Transforms and Applications Handbook*, 3rd ed. CRC Press, 2018.

[18] W. L. Briggs and V. E. Henson, *The DFT: An Owner's Manual for the Discrete Fourier Transform.* Society for Industrial and Applied Mathematics, 1995.

[19] M. Skolnik, *Radar Handbook*, 3rd ed. Mc Graw Hill, 2008.

[20] M. A. Richards, J. A. Scheer, and W. A. Holm, *Principles of Modern Radar, Vol I: Basic Principles.* Scitech Publishing, 2010.

[21] D. W. Bliss and K. W. Forsythe, "Multiple-input multiple-output (MIMO) radar and imaging: degrees of freedom and resolution," in *The Thrity-Seventh Asilomar Conference on Signals, Systems Computers, 2003*, vol. 1, 2003, pp. 54–59 Vol.1.

[22] T. Kilpatrick and I. D. Longstaff, "Generalising the co-array, for SAR and MIMO radar," in *2015 IEEE Radar Conference (RadarCon)*, 2015, pp. 1188–1192.

[23] M. A. Richards, *Fundamentals of Radar Signal Processing*, 2nd ed. McGraw-Hill Education, 2014.

[24] R. Rasshofer and G. K, "Automotive Radar and Lidar Systems for Next Generation Driver Assistance Functions," *Advances in Radio Science - Kleinheubacher Berichte*, vol. 3, May 2005.

[25] *Chassis Systems Control – Mittelbereichsradarsensor (MRR) für Front- und Heckanwendungen*, Robert Bosch GmbH, 2015. [Online]. Available: https://www.bosch-mobility-solutions.com/media/global/products-and-services/passenger-cars-and-light-commercial-vehicles/driver-assistance-systems/spurwechselassistent/mittelbereichsradarsensor-(mrr-rear)/produktdatenblatt-mittelbereichsradarsensor-mrr-rear.pdf

[26] F. de Ponte Müller, "Survey on ranging sensors and cooperative techniques for relative positioning of vehicles," *Sensors (Switzerland)*, vol. 17, no. 2, 2017.

[27] J. Hasch, E. Topak, R. Schnabel, T. Zwick, R. Weigel, and C. Waldschmidt, "Millimeter-Wave Technology for Automotive Radar Sensors in the 77 GHz Frequency Band," *IEEE Transactions on Microwave Theory and Techniques*, vol. 60, no. 3, pp. 845–860, 2012.

[28] M. Schneider, "Automotive Radar – Status and Trends," *2005 German Microwave Conference (GeMiC)*, 2005.

[29] M. Röckl, J. Gacnik, J. Schomerus, T. Strang, and M. Kranz, "Sensing the environment for future driver assistance combining autonomous and cooperative appliances," in *Fourth International Workshop on Vehicle-to-Vehicle Communications (V2VCOM)*, 2008, pp. 45–56.

[30] M. Mi, M. Moallem, J. Chen, M. Li, and R. Murugan, "Package Co-design of a Fully Integrated Multimode 76-81GHz 45nm RFCMOS FMCW Automotive Radar Transceiver," in *2018 IEEE 68th Electronic Components and Technology Conference (ECTC)*, 2018, pp. 1054–1061.

[31] *AWR1243 Single-Chip 77- and 79-GHz FMCW Transceiver*, Texas Instruments, 2019. [Online]. Available: https://www.ti.com/lit/ds/symlink/awr1243.pdf

[32] *Radar sensors for automotive*, Infineon, 2019. [Online]. Available: https://www.infineon.com/cms/en/product/sensor/radar-image-sensors/radar-sensors/radar-sensors-for-automotive/?redirId=119177

[33] *76 GHz to 81 GHz car RADAR transceiver*, NXP, 2019. [Online]. Available: https://www.nxp.com/docs/en/data-sheet/TEF810XDS.pdf

[34] J. C. Scheytt, Y. Sun, K. Schmalz, Y. Mao, R. Wang, W. Debski, and W. Winkler, "Towards mm-wave System-On-Chip with integrated antennas for low-cost 122 and 245 GHz radar sensors," in *2013 IEEE 13th Topical Meeting on Silicon Monolithic Integrated Circuits in RF Systems*, 2013, pp. 246–248.

[35] M. G. Girma, J. Hasch, I. Sarkas, S. P. Voinigescu, and T. Zwick, "122 GHz radar sensor based on a monostatic SiGe-BiCMOS IC with an on-chip antenna," in *2012 7th European Microwave Integrated Circuit Conference*, 2012, pp. 357–360.

[36] S. Yuan and H. Schumacher, "110–140-GHz single-chip reconfigurable radar frontend with on-chip antenna," in *2015 IEEE Bipolar/BiCMOS Circuits and Technology Meeting - BCTM*, 2015, pp. 48–51.

[37] S. S. Ahmed, A. Schiessl, and L. Schmidt, "A Novel Fully Electronic Active Real-Time Imager Based on a Planar Multistatic Sparse Array," *IEEE Transactions on Microwave Theory and Techniques*, vol. 59, no. 12, pp. 3567–3576, 2011.

[38] S. S. Ahmed, A. Schiessl, F. Gumbmann, M. Tiebout, S. Methfessel, and L. Schmidt, "Advanced Microwave Imaging," *IEEE Microwave Magazine*, vol. 13, no. 6, pp. 26–43, 2012.

[39] S. S. Ahmed, *Electronic Microwave Imaging with Planar Multistatic Arrays.* Logos Berlin, 2014.

[40] V. Bach, *Sicherheit durch Technik*, Rohde und Schwarz GmbH, 2016. [Online]. Available: https://cdn.rohde-schwarz.com/magazine/pdfs_1/article/216/german_13/FirstSpirit_1481140193631NEWS_216__QPS_german.pdf

[41] V. D. Agrawal and Y. T. Lo, "Distribution of sidelobe level in random arrays," *Proceedings of the IEEE*, vol. 57, no. 10, pp. 1764–1765, 1969.

[42] P. L. T. Bui, P. Rocca, and R. L. Haupt, "Aperiodic Planar Array Synthesis Using Pseudo-Random Sequences," in *2018 International Applied Computational Electromagnetics Society Symposium - China (ACES)*, 2018, pp. 1–2.

[43] R. E. Collin and F. J. Zucker, *Antenna Theory, Part 1*, 1st ed. McGraw-Hill, 1969.

[44] A. Moffet, "Minimum-redundancy linear arrays," *IEEE Transactions on Antennas and Propagation*, vol. 16, no. 2, pp. 172–175, 1968.

[45] K. A. Blanton and J. H. McClellan, "New search algorithm for minimum redundancy linear arrays," in *[Proceedings] ICASSP 91: 1991 International Conference on Acoustics, Speech, and Signal Processing*, 1991, pp. 1361–1364 vol.2.

[46] Ç. Korkuç and E. Afacan, "Optimization of circular antenna arrays using particle swarm optimization," in *2016 24th Signal Processing and Communication Application Conference (SIU)*, 2016, pp. 449–452.

[47] V. S. Gangwar, A. K. Singh, E. Thomas, and S. P. Singh, "Side lobe level suppression in a thinned linear antenna array using particle swarm optimization," in *2015 International Conference on Applied and Theoretical Computing and Communication Technology (iCATccT)*, 2015, pp. 787–790.

[48] E. R. Schlosser, S. M. Tolfo, and M. V. T. Heckler, "Particle Swarm Optimization for antenna arrays synthesis," in *2015 SBMO/IEEE MTT-S International Microwave and Optoelectronics Conference (IMOC)*, 2015, pp. 1–6.

[49] M. Ridwan, M. Abdo, and E. Jorswieck, "Design of non-uniform antenna arrays using genetic algorithm," in *13th International Conference on Advanced Communication Technology (ICACT2011)*, 2011, pp. 422–427.

[50] B. V. Ha, M. Mussetta, P. Pirinoli, and R. E. Zich, "Modified Compact Genetic Algorithm for Thinned Array Synthesis," *IEEE Antennas and Wireless Propagation Letters*, vol. 15, pp. 1105–1108, 2016.

[51] M. Gen and R. Cheng, *Genetic Algorithms and Engineering Optimization.* Wiley, 2000.

[52] A. Konak, D. W. Coit, and A. E. Smith, "Multi-objective optimization using genetic algorithms: A tutorial," *Reliability Engineering & System Safety*, vol. 91, no. 9, pp. 992 – 1007, 2006, special Issue - Genetic Algorithms and Reliability.

[53] M. Shibao, K. Uchiyama, and A. Kajiwara, "RCS Characteristics of Road Debris at 79GHz Millimeter-wave Radar," in *2019 IEEE Radio and Wireless Symposium (RWS)*, 2019, pp. 1–4.

[54] D. Belgiovane, C. Chen, M. Chen, S. Y. . Chien, and R. Sherony, "77 GHz radar scattering properties of pedestrians," in *2014 IEEE Radar Conference*, 2014, pp. 0735–0738.

[55] D. Marpaung, C. Roeloffzen, R. Heideman, A. Leinse, S. Sales, and J. Capmany, "Integrated microwave photonics," *Laser & Photonics Reviews*, vol. 7, no. 4, pp. 506–538, 2013.

[56] P. Ghelfi, F. Laghezza, F. Scotti, G. Serafino, A. Capria, S. Pinna, D. Onori, C. Porzi, M. Scaffardi, A. Malacarne, V. Vercesi, E. Lazzeri, F. Berizzi, and A. Bogoni, "A fully photonics-based coherent radar system," *Nature*, vol. 507, no. 7492, pp. 341–345, 2014.

[57] S. Preussler, F. Schwartau, J. Schoebel, and T. Schneider, "Photonically synchronized radar for advanced driver assistance systems," in *Terahertz, RF, Millimeter, and Submillimeter-Wave Technology and Applications XII*, vol. 10917, Mar. 2019, p. 9.

[58] S. Preussler, F. Schwartau, J. Schoebel, and T. Schneider, "Photonically synchronized large aperture radar for autonomous driving," *Optics Express*, vol. 27, p. 1199, Jan. 2019.

[59] S. Preussler, F. Schwartau, J. Schoebel, and T. Schneider, "Optical Signal Generation and Distribution for Large Aperture Radar in Autonomous Driving," in *2019 12th German Microwave Conference (GeMiC)*, Mar. 2019, pp. 154–157.

[60] F. Schwartau, S. Preussler, M. Krueckemeier, F. Pfeiffer, H. Stuelzebach, T. Schneider, and J. Schoebel, "Modular Wideband High Angular Resolution 79 GHz Radar System," in *2019 12th German Microwave Conference (GeMiC)*, Mar. 2019, pp. 194–197.

[61] *Rogers RO3000 Series Circuit Materials*, Rogers Corporation, 2019. [Online]. Available: https://rogerscorp.com/-/media/project/rogerscorp/documents/advanced-connectivity-solutions/english/data-sheets/ro3000-laminate-data-sheet-ro3003----ro3006----ro3010----ro3035.pdf

[62] B. Jian, J. Yuan, and Q. Liu, "Procedure to Design a Series-fed Microstrip Patch Antenna Array for 77 GHz Automotive Radar," in *2019 Cross Strait Quad-Regional Radio Science and Wireless Technology Conference (CSQRWC)*, 2019, pp. 1–2.

[63] F. Pfeiffer, Personal Communication, Perisens GmbH, September 2017.

[64] E. Fishler, A. Haimovich, R. Blum, D. Chizhik, L. Cimini, and R. Valenzuela, "MIMO radar: an idea whose time has come," in *Proceedings of the 2004 IEEE Radar Conference*, 2004, pp. 71–78.

[65] *AWR1642 Single-Chip 77- and 79-GHz FMCW Radar sensor*, Texas Instruments, 2020. [Online]. Available: https://www.ti.com/lit/ds/symlink/awr1642.pdf

[66] *AWR1642 Evaluation Module (AWR1642BOOST) Single-Chip mmWave Sensing Solution*, Texas Instruments, 2020. [Online]. Available: https://www.ti.com/lit/pdf/swru508

[67] A. Dürr, D. Schwarz, F. Roos, P. Hügler, S. Bucher, P. Grüner, and C. Waldschmidt, "On the Calibration of mm-Wave MIMO Radars Using Sparse Antenna Arrays for DoA Estimation," in *2019 16th European Radar Conference (EuRAD)*, 2019, pp. 349–352.

[68] J. H. Huang, J. L. Garry, G. E. Smith, and D. K. P. Tan, "In-field calibration of passive array receiver using detected target," in *2018 IEEE Radar Conference (RadarConf18)*, 2018, pp. 0715–0720.

[69] *Allgemeintoleranzen*, DIN Norm ISO 2768-1, Jun. 1991.

[70] W. M. Haynes, *CRC Handbook of Chemistry and Physics*, 92nd ed. CRC Press, 2011-2012.

[71] W. Kaiser, *Kunststoffchemie für Ingenieure.* Hanser, 2006.

[72] K. W. Kark, *Antennen und Strahlungsfelder*, 6th ed. Springer Vieweg, 2017.

List of Symbols

The following definitions are used, if not otherwise noted:

$a, a_{\mathrm{cr}} \in \mathbb{R}$	Corner reflector side edge
$a_{\mathrm{pl}} \in \mathbb{R}$	Plate reflector dimension (width/height)
$\boldsymbol{a}_r \in \mathbb{R}^3$	Location of the receive antenna r
$\boldsymbol{a}_t \in \mathbb{R}^3$	Kocation of the transmit antenna t
$\boldsymbol{a}_{t,r} \in \mathbb{R}^3$	Location of the virtual antenna of the transmitter t and receiver r (Mono-static if not otherwise noted)
$\boldsymbol{a}_{t,r,\mathrm{bs}} \in \mathbb{R}^3$	Location of the bi-static virtual antenna of the transmitter t and receiver r
$\boldsymbol{a}_{t,r,\mathrm{ms}} \in \mathbb{R}^3$	Location of the mono-static virtual antenna of the transmitter t and receiver r
$A_{t,r,\mathrm{exp}} \in \mathbb{R}$	Expected amplitude between transmitter and receiver
$A_{t,r,\mathrm{meas}} \in \mathbb{R}$	Measured amplitude between transmitter and receiver
$A_{t,r,\mathrm{err}} \in \mathbb{R}$	Amplitude error between transmitter and receiver
$\boldsymbol{a}_{u,v} \in \mathbb{R}^3$	Two-dimensional grid
$A_{\mathrm{w,mut}} \in \mathbb{R}$	Weight optimizer, mutation amplitude deviation
$B \in \mathbb{R} \in \mathbb{R}$	System or sweep bandwidth
$B_{\mathrm{bin}} \in \mathbb{R}$	Bandwidth of a single FFT bin
$\boldsymbol{B}^{\mathrm{ff}} \in \mathbb{C}^{N_{\mathrm{t}} \times N_{\mathrm{r}}}$	Beamforming matrix in farfield
$\boldsymbol{B}^{\mathrm{nf}} \in \mathbb{C}^{N_{\mathrm{t}} \times N_{\mathrm{r}}}$	Beamforming matrix in nearfield
$c_0 \in \mathbb{R}$	Speed of light in vacuum
$C_{t,r} \in \mathbb{C}$	Complex calibration term for transmitter and receiver
$\boldsymbol{c} \in \mathbb{R}^3$	Center of virtual array
$\boldsymbol{c}_q \in \mathbb{R}^3$	Center of quadrant q of virtual array
$D \in \mathbb{R}$	Aperture of an array
$D(\theta, \varphi) \in \mathbb{R}$	Directivity radiation pattern
$d \in \mathbb{R}$	Distance (usually to a target)
$D_0 \in \mathbb{R}$	Directivity of main lobe
$D_{\mathrm{array}} \in \mathbb{R}$	Transmit and receive array directivity
$D_{\mathrm{cr}} \in \mathbb{R}$	Aperture of a corner reflector
$D_{\mathrm{el}}(\theta, \varphi) \in \mathbb{R}$	Directivity radiation pattern of an array element
$D_{\mathrm{F}}(\theta, \varphi) \in \mathbb{R}$	Directivity of array factor
$d_{\mathrm{ff}} \in \mathbb{R}$	Far-field distance
$d_{\mathrm{max}} \in \mathbb{R}$	Maximum unambiguity distance
$D_{\mathrm{max}}(\theta, \varphi) \in \mathbb{R}$	Combined directivity of array and element
$d_{\mathrm{min}} \in \mathbb{R}$	Minimum distance of a target
$D_{\mathrm{r}} \in \mathbb{R}$	Receive aperture
$D_{\mathrm{t}} \in \mathbb{R}$	Transmitter aperture
$D_{\mathrm{tot}}(\theta, \varphi) \in \mathbb{R}$	Directivity radiation pattern of array including elements
$D_{\mathrm{d}} \in \mathbb{R}$	Maximum distance between any Tx/Rx combination

$E(r,\theta,\varphi) \in \mathbb{R}$	Field strength
$E_{\mathrm{el}}(r,\theta,\varphi) \in \mathbb{R}$	Field strength of array element
$E_{\mathrm{tot}}(r,\theta,\varphi) \in \mathbb{R}$	Field strength of array including the elements
$\boldsymbol{E}_{\mathrm{i},g} \in \mathbb{R}^3$	Actual module positions vector
$\boldsymbol{E}_{\mathrm{nm},o,g} \in \mathbb{R}^3$	Normalized estimated displacement vector for module g in iteration o
$\boldsymbol{E}_{o,g} \in \mathbb{R}^3$	Estimated displacement vector for module g in iteration o
$\mathrm{err} \in \mathbb{R}$	Relative RMS error for HPBW
$F \in \mathbb{R}$	Receiver's noise figure
$F(\theta,\varphi) \in \mathbb{C}$	Array factor with spherical coordinates
$F(\gamma,\delta) \in \mathbb{C}$	Array factor with azimuth/elevation
$f_{\mathrm{b}} \in \mathbb{R}$	Beat frequency without Doppler
$f_{\mathrm{c}} \in \mathbb{R}$	Center frequency
$f_{\mathrm{bd}} \in \mathbb{R}$	Beat frequency during down sweep
$f_{\mathrm{bu}} \in \mathbb{R}$	Beat frequency during up sweep
$f_{\mathrm{d}} \in \mathbb{R}$	Doppler frequency
$f_{\mathrm{d,max}} \in \mathbb{R}$	Maximum Doppler frequency
$f_{\mathrm{pos}} \in \mathbb{R}$	Position optimizer, fitness function
$f_{\mathrm{s,adc}} \in \mathbb{R}$	ADC sample rate
$g_{\delta,f_{\mathrm{d}},t,r} \in \mathbb{C}$	Element in range-Doppler matrix for Tx/Rx combination t/r
$g_{\mathrm{comp},\delta,f_{\mathrm{d}},t,r} \in \mathbb{C}$	Movement compensated element in range-Doppler matrix
$G(\theta,\varphi) \in \mathbb{R}$	Gain radiation pattern
$G_{\mathrm{el}}(\theta,\varphi) \in \mathbb{R}$	Gain radiation pattern of array element
$G_{\mathrm{r}} \in \mathbb{R}$	Overall receive gain
$G_{\mathrm{r,el}} \in \mathbb{R}$	Receive gain of a single element
$G_{\mathrm{t}} \in \mathbb{R}$	Overall transmit gain
$G_{\mathrm{tot}}(\theta,\varphi) \in \mathbb{R}$	Gain radiation pattern of array including elements
$G_{\mathrm{t,el}} \in \mathbb{R}$	Transmit gain of a single element
$h_{t,r} \in \mathbb{R}$	Sample for Tx/Rx combination t/r incl. beamforming
$h_{\mathrm{norm},t,r} \in \mathbb{R}$	Normalized sample for Tx/Rx combination t/r incl. beamforming
$h_{\mathrm{mod},g_{\mathrm{t}},g_{\mathrm{r}}} \in \mathbb{R}$	Averaged samples for for Tx/Rx module combination $g_{\mathrm{t}}/g_{\mathrm{r}}$ incl. beamforming
$\boldsymbol{H}_{\delta}^{\mathrm{ff}} \in \mathbb{C}^{N_{\mathrm{t}} \times N_{\mathrm{r}}}$	Channel response at delay δ in farfield
$\boldsymbol{H}_{\delta}^{\mathrm{nf}} \in \mathbb{C}^{N_{\mathrm{t}} \times N_{\mathrm{r}}}$	Channel response at delay δ in nearfield
$\mathrm{HPBW} \in \mathbb{R}$	Half power beam width
$\mathrm{HPBW}_{\mathrm{cr}} \in \mathbb{R}$	Half power beam width of a corner reflector
$\mathrm{HPBW}_{\mathrm{pl}} \in \mathbb{R}$	Half power beam width of a plate
$I \in \mathbb{R}$	Integral coefficient of PID controller
$k \in \mathbb{R}$	Wavenumber
$\boldsymbol{k} \in \mathbb{R}^3$	Wavenumber vector
$k_0 \in \mathbb{R}$	Boltzmann constant
$k_{\mathrm{red}} \in \mathbb{R}$	Position optimizer, redundancy penalty factor
$k_{\mathrm{round}} \in \mathbb{R}$	Position optimizer, roundness penalty factor
$L \in \mathbb{R}$	Corner reflector front edge length
$\mathrm{l_f}$	Patch feed width
$\mathrm{l_l}$	Patch length
$l_{\mathrm{max,t}} \in \mathbb{R}$	Maximum microstrip transmitter length
$l_{\mathrm{max,r}} \in \mathbb{R}$	Maximum microstrip receiver length
$L_{\mathrm{misc}} \in \mathbb{R}$	Miscellaneous losses

$\mathrm{l_s}$	Patch slot width
$\mathrm{l_t}$	Patch slot length
$L_{\mathrm{tot}} \in \mathbb{R}$	Additional system losses
$L_{\mathrm{us}} \in \mathbb{R}$	Microstrip losses per millimeter
$L_{\mathrm{us,t}} \in \mathbb{R}$	Maximum transmit microstrip losses
$L_{\mathrm{us,r}} \in \mathbb{R}$	Maximum receive microstrip losses
$\mathrm{l_w}$	Patch width
$m \in \mathbb{N}$	First output axis for 1D/2D iDFT
$M \in \mathbb{N}$	Number of iDFT elements in first axis
$n \in \mathbb{N}$	Second output axis for 2D iDFT
$N \in \mathbb{N} \in \mathbb{N}$	Number of iDFT elements in second axis
$N_{\mathrm{adc}} \in \mathbb{N}$	Number of ADC samples/samples per sweep
$N_{\mathrm{ft}} \in \mathbb{N}$	Number of samples per chirp (fast time elements)
$N_{\mathrm{g}} \in \mathbb{N}$	Number of modules in system
$N_{\mathrm{p,pop}} \in \mathbb{N}$	Position optimizer, population size
$N_{\mathrm{p,keep}} \in \mathbb{N}$	Position optimizer, chromosomes to keep
$N_{\mathrm{p,new}} \in \mathbb{N}$	Position optimizer, new chromosomes
$N_{\mathrm{p,gen}} \in \mathbb{N}$	Position optimizer, number of generations
$N_{\mathrm{r}} \in \mathbb{N}$	Number of receive antennas
$N_{\mathrm{st}} \in \mathbb{N}$	Number of chirps (slow time elements)
$N_{\mathrm{st,min}} \in \mathbb{N}$	Minimum number of chirps (slow time elements)
$N_{\mathrm{t}} \in \mathbb{N}$	Number of transmit antennas
$N_{\mathrm{train}} \in \mathbb{N}$	Number of training cells (CFAR)
$N_{\mathrm{w,pop}} \in \mathbb{N}$	Weight optimizer, population size
$N_{\mathrm{w,keep}} \in \mathbb{N}$	Weight optimizer, chromosomes to keep
$N_{\mathrm{w,new}} \in \mathbb{N}$	Weight optimizer, new chromosomes
$N_{\mathrm{w,gen}} \in \mathbb{N}$	Weight optimizer, number of generations
$\boldsymbol{p}_{d,\varphi,\theta} \in \mathbb{R}^3$	Target position or point of beamforming focus
$P_{\mathrm{FA}} \in \mathbb{R}$	False alarm rate (CFAR)
$P_{\mathrm{in,array}} \in \mathbb{R}$	Input power of array
$P_{\mathrm{in,el}} \in \mathbb{R}$	Input power of array element
$P_{\mathrm{n}} \in \mathbb{R}$	Noise power
$p_{\mathrm{p,mut}} \in \mathbb{R}$	Position optimizer, mutation probability
$p_{\mathrm{p,repr}} \in \mathbb{R}$	Position optimizer, reproduction ratio
$\boldsymbol{p}_{\varphi,\theta} \in \mathbb{R}^3$	Target or beamforming direction of unit length
$P_{\mathrm{rad}} \in \mathbb{R}$	Radiated power
$P_{\mathrm{rad,el}} \in \mathbb{R}$	Radiated power of an array element
$P_{\mathrm{rad,array}} \in \mathbb{R}$	Radiated power of array with isotropic elements
$P_{\mathrm{r,min}} \in \mathbb{R}$	Minimum receive power
$P_{\mathrm{red}} \in \mathbb{R}$	Position optimizer, redundancy penalty
$P_{\mathrm{round}} \in \mathbb{R}$	Position optimizer, roundness penalty
$P_{\mathrm{t}} \in \mathbb{R}$	Overall transmit power
$P_{\mathrm{t,el}} \in \mathbb{R}$	Transmit power of a single element
$p_{\mathrm{w,mut}} \in \mathbb{R}$	Weight optimizer, mutation probability
$p_{\mathrm{w,repr}} \in \mathbb{R}$	Weight optimizer, reproduction ratio
$q \in \mathbb{N}$	Quadrant number
$r \in \mathbb{N}$	Index of receive antenna
$s \in \mathbb{R}$	Frequency slope
$\mathrm{SLL} \in \mathbb{R}$	Side lobe level
$\mathrm{SNR_{min}} \in \mathbb{R}$	Minimum signal-to-noise ratio

$t \in \mathbb{N}$	Index of transmit antenna
$t \in \mathbb{R}$	Time
$T_0 \in \mathbb{R}$	System temperature
$t_{\text{int}} \in \mathbb{R}$	Total integration time
$t_{\text{s}} \in \mathbb{R}$	Duration of a single sweep or ramp
$u \in \mathbb{N}$	First input axis for 1D/2D iDFT
$U(\theta, \varphi) \in \mathbb{R}$	Radiation intensity
$U_{\text{bias,new}} \in \mathbb{R}$	New bias voltage of PID controller
$U_{\text{bias,old}} \in \mathbb{R}$	Last bias voltage of PID controller
$U_{\text{el}}(\theta, \varphi) \in \mathbb{R}$	Radiation intensity of array element
$U_{\text{mon}} \in \mathbb{R}$	Measured monitor voltage of PID controller
$U_{\text{mon,tgt}} \in \mathbb{R}$	Target monitor voltage of PID controller
$U_{\text{tot}}(\theta, \varphi) \in \mathbb{R}$	Radiation intensity of array including elements
$T_{\text{s}} \in \mathbb{R}$	Period of sweeps
$v \in \mathbb{N}$	Second input axis for 2D iDFT
$v_{\text{r}} \in \mathbb{R}$	Radial target velocity
$v_{\text{r,max}} \in \mathbb{R}$	Maximum radial target velocity
$\boldsymbol{w}_i \in \mathbb{R}^{N_{\text{t}} N_{\text{r}}}$	Weight vector of a virtual array
$\boldsymbol{w}_t \in \mathbb{R}^{N_{\text{t}}}$	Weight vector of a transmit array
$\boldsymbol{w}_r \in \mathbb{R}^{N_{\text{r}}}$	Weight vector of a receive array
$Z_0 \in \mathbb{R}$	Intrinsic impedance of a medium
$X(u), X(u, v) \in \mathbb{C}$	1D and 2D iDFT input
$x(m), x(m, n) \in \mathbb{C}$	1D and 2D iDFT result

$\alpha \in \mathbb{R}$	CFAR threshold factor
$\beta \in \mathbb{R}$	Scaling factor for module displacement optimization
$\gamma \in \mathbb{R}$	Azimuth
$\gamma_{\text{tgt}} \in \mathbb{R}$	Azimuth of a target
$\delta \in \mathbb{R}$	Elevation
$\delta_{\text{tgt}} \in \mathbb{R}$	Elevation of a target
$\Delta d_{g_{\text{t}},g_{\text{r}}} \in \mathbb{R}$	Distance offset in module displacement compensation
$\Delta \boldsymbol{E}_{g_{\text{t}},g_{\text{r}}} \in \mathbb{R}^3$	Module displacement correction vector
$\Delta_{\text{max}} \in \mathbb{R}$	Maximum delay between any Tx/Rx path
$\Delta R \in \mathbb{R}$	Range resolution
$\Delta R_{\text{h}} \in \mathbb{R}$	Range resolution, with window
$\Delta R_{\text{h,meas}} \in \mathbb{R}$	Range resolution, with window, measured
$\Delta l_{\text{max}} \in \mathbb{R}$	Length equivalent to ΔT_{max}
$\Delta T_{\text{max}} \in \mathbb{R}$	Maximum delay difference between any path
$\Delta l_{\text{el}} \in \mathbb{R}$	Path length difference in electrical wires
$\Delta l_{\text{path}} \in \mathbb{R}$	Path length difference due to Tx/Rx positions
$\Delta l_{\text{sync}} \in \mathbb{R}$	Path length difference in synchronization system
$\Delta l_{\text{wc}} \in \mathbb{R}$	Worst case path length difference
$\Delta \gamma_{\text{sim}} \in \mathbb{R}$	Azimuth resolution, simulated
$\Delta \gamma_{\text{meas}} \in \mathbb{R}$	Azimuth resolution, measured
$\Delta \delta_{\text{sim}} \in \mathbb{R}$	Elevation resolution, simulated
$\Delta \delta_{\text{meas}} \in \mathbb{R}$	Elevation resolution, measured
$\Delta \Phi_t \in \mathbb{R}$	Phase offset caused by target movement
$\epsilon_o \in \mathbb{R}$	RMS error in estimated module position offset
$\Theta \in \mathbb{R}$	Angle factor for HPBW estimation of a corner reflector
$\theta, \varphi \in \mathbb{R}$	Angles of target or beamforming point in spherical coordinates
$\hat{\theta}, \hat{\varphi} \in \mathbb{R}$	Angles of spherical coordinates for integrals
$\theta_0, \varphi_0 \in \mathbb{R}$	Direction of main lobe
$\lambda \in \mathbb{R}$	Wavelength
$\nu \in \mathbb{R}$	Array power gain factor
$\sigma \in \mathbb{R}$	Radar cross section
$\sigma_{\text{cr}} \in \mathbb{R}$	Radar cross section of a corner reflector
$\sigma_{\text{pl}} \in \mathbb{R}$	Radar cross section of a plate
$\tau \in \mathbb{R}$	Reflected signal delay
$\phi(t) \in \mathbb{R}$	Phase of target reflection over time
$\phi_{t,r} \in \mathbb{R}$	Phase from transmitter to receiver
$\phi_{t,r,\text{exp}} \in \mathbb{R}$	Expected phase from transmitter to receiver
$\phi_{t,r,\text{meas}} \in \mathbb{R}$	Measured phase from transmitter to receiver
$\phi_{t,r,\text{offs}} \in \mathbb{R}$	Phase offset from transmitter to receiver
$\omega_{\text{d}} \in \mathbb{R}$	Doppler angular frequency

List of Publications

- **Fabian Schwartau**, Yannic Schröder, Lars Wolf and Joerg Schoebel, "*Large Minimum Redundancy Linear Arrays: Systematic Search of Perfect and Optimal Rulers Exploiting Parallel Processing*," IEEE Open Journal of Antennas and Propagation (Early Access) 2020, accepted with minor revisions
- **Fabian Schwartau**, Carsten Monka-Ewe, Reinhard Caspary, Wolfgang Kowalsky and Joerg Schoebel, "*A Two-Dimensional Continuous-Wave Imaging System for Scanning of Dielectric Substrates at Millimeter-Wave Frequencies*," Kleinheubacher Tagung 2019 (KH 2019), Miltenberg, Germany
- **Fabian Schwartau**, Stefan Preussler, Markus Krueckemeier, Florian Pfeiffer, Hannes Stuelzebach, Thomas Schneider and Joerg Schoebel, "*Modular Wideband High Angular Resolution 79 GHz Radar System*," 12th German Microwave Conference (GeMiC 2019), Stuttgart, Germany
- **Fabian Schwartau**, Carsten Monka-Ewe Markus Krueckemeier and Joerg Schoebel, "*Aircraft window attenuation measurements at 60 GHz for wireless in-cabin communication*," 11th German Microwave Conference (GeMiC 2018), Freiburg, Germany
- Stefan Preussler, **Fabian Schwartau**, Joerg Schoebel and Thomas Schneider, "*Photonic Components for Signal Generation and Distribution for Large Aperture Radar in Autonomous Driving*," Frequenz 2019
- Markus Krueckemeier, **Fabian Schwartau** and Joerg Schoebel, "*A modular localization system combining passive RF detection and passive radar*," Kleinheubacher Tagung 2019 (KH 2019), Miltenberg, Germany
- Markus Krueckemeier, **Fabian Schwartau**, Carsten Monka-Ewe and Joerg Schoebel, "*Synchronization of Multiple USRP SDRs for Coherent Receiver Applications*," 6th International Conference on Software Defined Systems (SDS 2019), Rome, Italy
- Stefan Preussler, **Fabian Schwartau**, Thomas Schneider and Joerg Schoebel, "*Optical Signal Generation and Distribution for Large Aperture Radar in Autonomous Driving*," 12th German Microwave Conference (GeMiC 2019), Stuttgart, Germany
- Stefan Preussler, **Fabian Schwartau**, Thomas Schneider and Joerg Schoebel, "*Photonically synchronized large aperture radar for autonomous driving*," Optics Express 2019 27:2, pp. 1199-1207
- Stefan Preussler, **Fabian Schwartau**, Thomas Schneider and Joerg Schoebel, "*Photonically synchronized radar for advanced driver assistance systems*," Proceedings of SPIE - The International Society for Optical Engineering (SPIE 2019)
- Björn Gernert, **Fabian Schwartau**, Sebastian Raabe, Joerg Schoebel, Karsten Schubert, Stephan Rottmann and Lars C. Wolf, "*Evaluation of suitable radio frequencies for data transmission in potato warehouses*," 15th International Conference on Mobile Ad Hoc and Sensor Systems (MASS 2018), Sheffield, United Kingdom
- Karsten Schubert, **Fabian Schwartau** and Jens Werner, "*Direction-dependent interference effect of a wind turbine on C-band weather radars*," Internationale Fachmesse und Kongress für Elektromagnetische Verträglichkeit (EMV 2018), Düsseldorf, Germany
- Karsten Schubert, **Fabian Schwartau** and Jens Werner, "*Experimentelles FMCW-Radar zur hochfrequenten Charakterisierung von Windenergieanlagen*," Advances in Radio Science 2017 15:1-9

- Karsten Schubert, **Fabian Schwartau** and Jens Werner, "*Design of an bi-static C-band radar for characterisation of wind turbine plants*," 2017 Asia-Pacific International Symposium on Electromagnetic Compatibility, (APEMC 2017), Seoul, Korea
- Edson P. da Silva, Robert Borkowski, Stefan Preussler, **Fabian Schwartau**, Simone Gaiarin, Miguel I. Olmedo, Armand Vedadi, Molly Piels, Michael Galili, Pengyu Guan, Sergei Popov, Camille-Sophie Brès, Thomas Schneider, Leif K. Oxenløwe, Darko Zibar, "*Combined Optical and Electrical Spectrum Shaping for High-Baud-Rate Nyquist-WDM Transceivers*," IEEE Photonics Journal 2016 8:1

Papers published after submission

- Sebastian Paul, **Fabian Schwartau**, Markus Krueckemeier, Reinhard Caspary, Joerg Schoebel, and Wolfgang Kowalsky, "*A Systematic Comparison of Near-Field Beamforming and Fourier-based Backward-Wave Holographic Imaging*," IEEE Open Journal of Antennas and Propagation 2021 2, pp. 921-931
- Markus Krueckemeier, **Fabian Schwartau**, Sebastian Paul and Joerg Schoebel, "*Transmitter Localization in Passive Radar and a Planar Approximation Method*," IEEE Transactions on Aerospace and Electronic Systems 2021 57:5, pp. 3405-3415

www.ingramcontent.com/pod-product-compliance
Ingram Content Group UK Ltd.
Pitfield, Milton Keynes, MK11 3LW, UK
UKHW022000190726
13853UKWH00004B/1644

9 783736 975071